Lecture Notes in Computer Science 12410

More information about this series at http://www.springer.com/series/8637

Abdelkader Hameurlain ·
A Min Tjoa (Eds.)

Transactions on Large-Scale Data- and Knowledge-Centered Systems XLVI

Springer

Editors-in-Chief
Abdelkader Hameurlain
IRIT, Paul Sabatier University
Toulouse, France

A Min Tjoa
IFS, Vienna University of Technology
Vienna, Austria

ISSN 0302-9743   ISSN 1611-3349   (electronic)
Lecture Notes in Computer Science
ISSN 1869-1994   ISSN 2510-4942   (electronic)
Transactions on Large-Scale Data- and Knowledge-Centered Systems
ISBN 978-3-662-62385-5   ISBN 978-3-662-62386-2   (eBook)
https://doi.org/10.1007/978-3-662-62386-2

This Springer imprint is published by the registered company Springer-Verlag GmbH, DE
part of Springer Nature
The registered company address is: Heidelberger Platz 3, 14197 Berlin, Germany

Preface

This volume contains six fully revised selected regular papers, covering a wide range of very hot topics in the fields of scalable and elastic framework for genomic data management, medical data cloud federations, mining sequential frequent patterns, scalable schema discovery, load shedding, and selectivity estimation using linked Bayesian networks.

We would like to sincerely thank the Editorial Board for thoroughly refereeing the submitted papers and ensuring the high quality of this volume. Special thanks go to Gabriela Wagner for her availability and her valuable work in the realization of this TLDKS volume.

August 2020   Abdelkader Hameurlain
                                                         A Min Tjoa

Organization

Editors-in-Chief

Abdelkader Hameurlain   Paul Sabatier University, IRIT, France
A Min Tjoa   Technical University of Vienna, IFS, Austria

Editorial Board

Reza Akbarinia   Inria, France
Dagmar Auer   FAW, Austria
Djamal Benslimane   Claude Bernard University Lyon 1, France
Stéphane Bressan   National University of Singapore, Singapore
Mirel Cosulschi   University of Craiova, Romania
Dirk Draheim   Tallinn University of Technology, Estonia
Johann Eder   University of Klagenfurt (AAU), Austria
Anastasios Gounaris   Aristotle University of Thessaloniki, Greece
Theo Härder   Technical University of Kaiserslautern, Germany
Sergio Ilarri   University of Zaragoza, Spain
Petar Jovanovic   Universitat Politècnica de Catalunya, BarcelonaTech, Spain
Aida Kamišalić Latifić   University of Maribor, Slovenia
Dieter Kranzlmüller   Ludwig-Maximilians-Universität München, Germany
Philippe Lamarre   INSA Lyon, France
Lenka Lhotská   Technical University of Prague, Czech Republic
Vladimir Marik   Technical University of Prague, Czech Republic
Jorge Martinez Gil   Software Competence Center Hagenberg, Austria
Franck Morvan   Paul Sabatier University, IRIT, France
Torben Bach Pedersen   Aalborg University, Denmark
Günther Pernul   University of Regensburg, Germany
Soror Sahri   Paris Descartes University, LIPADE, France
Shaoyi Yin   Paul Sabatier University, France
Feng (George) Yu   Youngstown State University, USA

Contents

Extracting Insights: A Data Centre Architecture Approach in Million Genome Era

Tariq Abdullah[(⊠)] and Ahmed Ahmet

University of Derby, Derby, UK
t.abdullah@derby.ac.uk, a.ahmet1@unimail.derby.ac.uk

Abstract. Advances in high throughput sequencing technologies have resulted in a drastic reduction in genome sequencing price and led to an exponential growth in the generation of genomic sequencing data. The genomics data is often stored on shared repositories and is both heterogeneous and unstructured in nature. It is both technically and culturally residing in big data domain due to the challenges of volume, velocity and variety.

Appropriate data storage and management, processing and analytic models are required to meet the growing challenges of genomic and clinical data. Existing research on the storage, management and analyses of genomic and clinical data do not provide a comprehensive solution, either providing Hadoop based solution lacking a robust computing solution for data mining and knowledge discovery, or a distributed in memory solution that are effective in reducing runtime but lack robustness on data store, resource management, reservation, and scheduling.

In this paper, we present a scalable and elastic framework for genomic data storage, management, and processing that addresses the weaknesses of existing approaches. Fundamental to our framework is a distributed resource management system with a plug and play NoSQL component and an in-memory, distributed computing framework with machine learning and visualisation plugin tools. We evaluated Avro, CSV, HBase, ORC, Parquet datastores and benchmark their performance. A case study of machine learning based genotype clustering is presented to demonstrate and evaluate the effectiveness of the presented framework. The results show an overall performance improvement of the genomics data analysis pipeline by 49% from existing approaches. Finally, we make recommendations on the state of the art technology and tools for effective architecture approaches for the management and knowledge discovery from large datasets.

1 Introduction

Since the completion of the Human Genomic Project in 2003, genomic and clinical data have seen continuous and unsustainable growth [3]. The human genome project took 10 years and a budget of $2.7 billion to be completed. Advances in

© Springer-Verlag GmbH Germany, part of Springer Nature 2020
A. Hameurlain and A M. Tjoa (Eds.): TLDKS XLVI, LNCS 12410, pp. 1–31, 2020.
https://doi.org/10.1007/978-3-662-62386-2_1

Next-Generation Sequencing (NGS) technologies have led to an unprecedented generation of genomics data [33,48,56]. NGS is a massively parallel genome sequencing technology delivering high throughput genome sequences (a generic NGS genome sequencing pipeline is depicted in Fig. 1). WGS and NGS sequencing technologies enable the inclusion of rare or even somatic mutations in the analysis, thus increasing the feature space by orders of magnitude and require a massively parallel approach to genomics data processing [53]. These breakthroughs in high-throughput sequencing over low-throughput sequencing, such as Sanger sequencing, have led to the price of sequencing to decrease dramatically[1]. The cost of Sequencing Genome illustrates how the adoption of NGS technology, starting in January 2008, led to an impressive drop in cost per genome. Before NGS technology [33], Sangar sequencing (first generation) technology was widely used[2]. Comparing the price trend to Moore's law we can see how the bottleneck in DNA analysis has shifted from sequencing to storage and processing of generated data. European Bioinformatics Institute (EBI) genomics data repository has a doubling time of 12 months across all data resources [10]. As of September 2017, EBI capacity stood at 120 Petabytes, they anticipate this figure to reach exabytes by 2022. Researchers are no longer dependent on their research laboratories for accessing genomic data. They rely on continuously growing data made available from various institutions on the cloud [9,10,16]. This alleviates constraints on local computing infrastructure. The challenges associated with genomic data generation put the management and analysis of genomic data in the big data domain.

WGS (Whole Genome Sequencing) technology has transformed the process of recording an individual's genetic code from a decade-long, billion-dollar, global endeavour to a week-long, $1000 service [12]. A single genome can vary from 700 MB to 200 GB. It is estimated that between 100 *million* and 2 *billion* human genomes could have been sequenced and 2–40 exabytes of data storage would be required. The computing resources for storing and processing this data will soon exceed those of Twitter and YouTube [7]. The 1000 genome project has deposited two times more raw data into NCBI's GenBank during its first 6 months than all the previous sequences deposited in the last 30 years [53]. This exponential growth in genomics data has resulted in an extraordinary volume of data generated at an unprecedented velocity from current and ongoing genomic projects [57]. The volume of genomics data is expected soon overtake social media, astronomy and social science [54]. Genomic data is mostly unstructured, recorded in a variety of data formats and require specialized platforms for efficiently managing, extracting insights, knowledge discovery and unlocking value from the deluge of genomic data are some of the key challenges for this domain.

The design and development of efficient and scalable computational platforms for meeting the above-mentioned challenges have lagged behind our ability to generate data [57]. A number of different approaches have been investigated that range from highly connected, customized server-based solutions [26]

---

[1] https://www.sevenbridges.com/rabixbeta/.
[2] www.sanger.ac.uk.

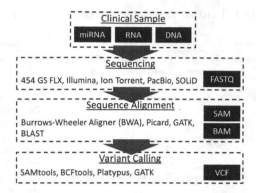

Fig. 1. NGS genomic data generation pipeline

to open-source Apache tools based platforms such as HDFS storage and computing architectures [14,30,35,39,43,49,50]. These studies focused on analysing different aspect of genomics data like variant annotation [39], alignment [43,50] genomics data quality control [49], general workflow management [51] and single nucleotide polymorphism calling [30]. They aimed to significantly speed up the batch-oriented analytic pipeline.

High-throughput sequencing has led to Genome-wide Association Studies (GWAS) that enabled researchers to determine the locations of variants between individuals [38]. The likelihood of groundbreaking discoveries in medical science and new medical insights like identification of unique disease contributing variants is now a real prospect [8,15].

Recent trends on the availability and reduced price of high-performance computing (HPC) have enabled new efficient in-memory computing frameworks to displace previously dominated batch-oriented frameworks for compute-intensive genomics data analytics. In in-memory approaches [4,11,36], an analytic pipeline reads the whole/part of genomics dataset (structured or unstructured) from storage for each analytic cycle. Reading data in this way is counterproductive and makes genomics data analysis inefficient with increased runtime and disk I/O bottlenecks. This overhead is a further complicating factor as genomic data is available in a compressed format. Compression pipeline parallelization [22] may relieve some of the overheads but does not address the inefficiency of reading data from storage every time.

To efficiently carry out knowledge discovery in genomics data, such as personalized healthcare, a platform requires a data warehouse with the ability to aggregate a multitude of structured and unstructured data that can be processed and analysed for value, delivering a complete medical view of the patient. This may include biological traits, environmental exposures and behaviours of each patient. However, this data may differ in nature, require different storage schema and distributed across diverse storage silos in medical health and bioinformatic organizations. These heterogeneous datasets must be integrated into a storage server for efficient data storage management and analysis.

Table 1. Existing studies for genomics data analysis

Study	Infrastructure	Strengths	Weaknesses
AzureBLAST	Windows Azure, BLAST, Azure Queues	Parallel computing, Distributed messaging	No in-memory computing, No storage infrastructure, No visualisation support
CloudBurst	Amazon's EC2, Apache Hadoop	Parallel computing	No in-memory computing, Distributed Storage, No visualisation support
BioPig	MapReduce, Apache Pig	Parallel Computing, High-level data flow abstraction	No in-memory computing, No storage infrastructure, No visualisation support
Bowtie aligner	Amazon EC2, Amazon S3	Parallel computing	No in-memory computing, No storage infrastructure, No visualisation support
Crossbow	Amazon Web Services, Apache Hadoop, Bowtie, SOAPsnp	Parallel computing, Distributed storage	No in-memory computing, No visualisation support
VariantSpark	Apache Spark, Spark's MLlib, Apache Hadoop	In-memory computing, Parallel Computing	No storage infrastructure, No visualisation support
SparkSeq	Apache Spark	In-memory computing, Parallel computing	No storage infrastructure, Lack of visualisation
ADAM	Apache Spark, Apache Parquet	In-memory computing, Parallel computing, SQL query engine	No visualisation support, No distributed storage
GenAp	Apache Spark, Spark SQL	In-memory computing, Parallel computing, Distributed SQL engine	No analytic support, No visualisation support
SeqPig	Apache Hadoop (MapReduce), Apache Pig	Parallel computing, High-level data flow abstraction	No in-memory computing, No storage infrastructure, No visualisation support

This work is an extension of our work [1] that proposed an in-memory computing approach in combination with a stateful storage infrastructure [Apache HBase] to meet the challenges of analysing genomics data. In this paper, we first present a scalable, elastic framework for genomic data storage, management, and processing for reducing the inefficient data read and disk I/O bottlenecks. Fundamental to this framework is a distributed resource management system with a plug and play NoSQL database components and is optimized for genomic data storage and retrieval. Secondly, the framework provides a distributed, elastic and intelligent in-memory computing approach that can integrate plug and play machine learning libraries and visualisation tools. Thirdly, we evaluate and benchmark the performance of the leading NoSQL databases (Apache Avro, HBase, ORC and Parquet datastores) on a variety of tasks which tests the storage and analytic components. Fourthly, a case study of machine learning based genotype clustering is presented to demonstrate and evaluate the effectiveness of our distributed, elastic, intelligent in-memory computing approach. Lastly, we make recommendations on tools and technologies for effective management of genomic data, knowledge discovery and extracting insights from it.

The paper is organized as follows: Sect. 2 provides a review of the state of the art genomics data storage, retrieval and processing approaches and a summary of issues and research gaps. The proposed framework and its components are explained in Sect. 3. A case study, "Genomics variant analysis", for evaluating the presented framework is explained in Sect. 3.4. A detailed discussion of the experimental results is provided in Sect. 4. This section also details the experimental setup and the dataset used for generating these results. The paper is concluded in Sect. 5 with future works.

2 Literature Review

The literature review focuses on identifying the gaps in the existing approaches for genomics data storage, retrieval and analysis pipelines. Due to the substantial drop in the sequencing cost, it is now economical to generate studies with cohort sizes previously reserved for larger consortia such as the 1000 genome project [9]. Many genomic data analysis platforms [29,30,32,37,50,52] are mostly batch processing systems, optimized for one pass batch processing without any support for interactive and/or ad hoc querying on the genomics data. CloudBurst [50] employs MapReduce for mapping single-end next-generation sequencing data to reference genomes. AzureBLAST [32] is a parallel Blast engine on the Windows Azure cloud platform for finding regions of local similarity between sequences[3]. BioPig [37] performs sequence analysis on large scale sequencing datasets using Hadoop and Apache Pig. Crossbow [30] performs human genome alignment and single nucleotide polymorphism detection on Hadoop based cluster. Bowtie aligner [29] aligns short DNA sequence reads to the human genome.

MapReduce based approaches transform data into 'key-value pairs' that can then be distributed between multiple nodes across a commodity computer cluster according to the size of a problem and these approaches are widespread in bioinformatics [19,21,25,30,46,47,50,55]. This is especially the case for sequence analysis tasks, such as read mapping [50], duplicate removal [25] and variant calling [30] as well as genome-wide analysis study based tasks [19,21]. Unfortunately, the MapReduce paradigm is not always the optimal solution, specifically for bioinformatics or machine learning applications that require iterative in-memory computation. Specifically, Hadoop is relying extensively on hard disk input-output operations (Disk IO), and this has proven to be a bottleneck in processing speed. Some studies [4,28,36,52] attempted to reduce or eliminate disk I/O operations by keeping data in memory. VariantSpark [4], performed a range of genome-based analysis tasks on VCF files, applied the K-means clustering algorithm for determining population structure from the 1000 genome project. SparkSeq [36] was developed for high-throughput sequence data analysis and supports filtering of reads, summarizing genomics features and statistical analysis using ADAM (a general purposes genomics framework and a set of formats and APIs as well as processing stage implementations for genomics data) [34]. While the speedup of ADAM over traditional methods was impressive (50

---

[3] http://blast.ncbi.nlm.nih.gov/Blast.cgi.

fold speedup), being limited by constraints within this general genomics framework can hamper performance. GenAp [28] provides a distributed SQL interface for genomics data by modifying Spark SQL. SeqPig [52] is a set of scripts for automatically generating MapReduce job for large sequencing datasets and uses Hadoop-BAM for reading input BAM files. Data provenance of bioinformatics workflows on PROV-DM model for re-executing the workflows in a more planned and controlled way [13]. Whereas, PostgresSQL [41] and Neo4J [42] are used for storing provenance data about the workflow execution for bioinformatics workflows, the raw data is not stored in both these approaches. NoSQL datastore [2] and document-oriented NoSQL datastore [17] to store and maintain persistency of genomics data. All these approaches lack a strategy for conceptually representing the data model associated with NoSQL datastores [40]. These approaches are summarised in Table 1.

Majority of the surveyed approaches read data from disk, try to optimally analyse the data in each read using a distributed, in-memory computing framework and lack a distributed data management/storage solution or visualisation platform simultaneously [14,30,35,39,43,49,50]. Some of the surveyed approaches do address computing and storage challenges, however, they don't have visualisation tools and plug-and-play machine learning libraries. Furthermore, these approaches don't consider the pre-processing of genomics data due to the unstructured nature of the genomics data and, therefore, result in further runtime overhead.

2.1   Genomics Data File Formats

The generation and processing of genomics data is generally classified into four categories: sequence data, annotations data, quantitative data and read alignments. Each of these genomics data categories has several specialised file formats (refer Table 2). Sequence data represents the nucleotide sequence of a chromosome, contig or transcript. Annotations data is the identification of features such as genes, SNPs, transcripts, with their locations of individual genes of sequences. Quantitative data are values associated with the chromosomal position. Read alignments are records where the sequence's place in the genome has been found using alignment tools. Table 2 shows a list of most widely used file formats in their aforementioned categories.

Figure 1 illustrates the key stages of NGS genomics data generation pipeline, from initial clinical samples to Variant Call Format (VCF) data and the utilization of some commonly used tools and file formats. Genomic sequencing and analysis pipelines comprise of sequencing, sequence alignment and variant calling stages. The process starts with samples of genetic information (like DNA, RNA or miRNA). Sequencing phase produces information containing fragments of DNA sequences, sequence letters with quality score. The Sequence alignment phase involves an aligner which is used to align the sequence fragments with the reference genome. Alignment phase can be broken down into two steps: 1) indexing the reference genome 2) Aligning the reads to the reference genome. Sequencing and Sequence alignment produces a DNA sequence for an individual

Table 2. Genomics data file formats

Annotations	Read alignment	Sequence data	Quantitative data
BED	Axt	BAM	bedGraph
bigBed	bigPsl	SAM	barChart
bigGenePred	bigMaf	CRAM	bigBarChart
bigNarrowPeak	bigChain	FASTA	bigWig
GFF	Chain	FASTAQ	GenePred
GTF	HAL	EMBL	WIG
Interact	MAF	GCG-RSF	
BigInteract	Net	GenBank	
	PSL	VCF	

but modern analysis pipelines don't stop there, the final phase, called Variant Calling, involves producing a set of variant calls for the individuals. This involves finding variants in the individual's sequence compared to the reference. The human genome comprises approximately 3 billion base-pairs of which 99.9% are similar. A single human genome sequenced at $30x$ coverage would produce ~200 GB of data. A variant file containing only a list of mutations would include approximately 3 million variants which would equal to ~130 MB in size [9].

The field of genomics is still evolving and a genomics file format convention is not yet established. The most widely used formats for sequencing data are tab-separated text files like FASTQ, FASTA, BAM, SAM and VCF [18,23,45].

3   Proposed Framework

This section outlines the proposed framework, its different components and the interaction between them (refer Fig. 3). The proposed framework provides a cloud-based, scalable, elastic genomic data storage, management, and processing for reducing the inefficient data read and disk I/O bottlenecks with a plug and play NoSQL database component that is optimized for genomic data storage and retrieval.

By placing genomics data files in a cloud-based distributed storage, we can perform parallel operations on the genomics data, keep intermediate results in-memory, query the genomics data without processing each time and can remove unnecessary disk I/O from the analytic pipeline. The in-memory intermediate results allow data to be requested much more quickly, rather than sending data back to the disk and making new requests. This in-memory approach, coupled with data parallelism, allows vast amounts of data to be processed in a significantly improved manner when compared to traditional distributed computing approaches (refer Sect. 4).

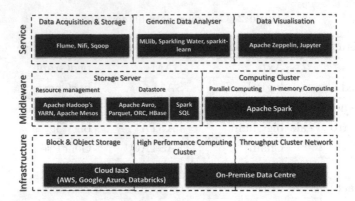

Fig. 2. Proposed framework

The framework presented here significantly extends our previous work [1] - that stored and analysed genomic data in one format only. It offers a complete data warehousing solution for a scalable and elastic data centre architecture and can store and analyse different genomics data formats (refer Sect. 2.1). It also has a distributed, intelligent in-memory computing component that can integrate plug-and-play machine learning libraries and visualisation tools for interactive data analytics visual tool.

The framework makes genomics data analysis efficient and reduces the processing latencies by transforming unstructured genomics data into a structured format removing the need to transform data with each new analytic task. This combined with distributed in-memory computing and reduces disk I/O bottlenecks. This also enables to aggregate a multitude of structured and unstructured data that can be processed and analysed for value, delivering a complete medical view of the patient. This may include biological traits, environmental exposures and behaviours of each patient. It also enables the organizations managing that lack the scalable, elastic storage and management system to keep up with rapidly expanding genomic data volumes and types.

Our framework is a distributed, scalable, fault-tolerant approach for storage, management and efficient analysis of genomics data. It's capable of handling both structured and unstructured genomics data which make up the full range of common biological (genomics) data types. In the remainder of this section, we explain the following three layers of our framework and interaction between them (refer Fig. 2)

1. Data Centre Services
2. Data Centre Middleware
3. Data Centre Infrastructure

By placing genomics data files in a cloud-based distributed storage, we can perform parallel operations and keep intermediate results in-memory while also query the unstructured genomics data without processing it each time, this removes unnecessary disk I/O from analytic pipeline. Intermediate results are kept in-memory to allow data to be requested much more quickly, rather than sending data back to disk and making new requests. This in-memory approach coupled with data parallelism allows vast amounts of data to be processed in a significantly improved manner when compared to traditional distributed computing approaches (refer Sect. 4).

3.1 Data Centre Services

This layer represents the services available to the user and is the point of interaction between user and framework. This is the top layer of our framework and provides three key services that enable the efficient storage, management and analysis of genomics data: 1) Data acquisition and Storage, 2) Data Analysis and 3) Data visualisation.

Data Acquisition
Genomics data comes in many formats for the various data types (refer Sect. 2.1). These are usually large plain text files and are available as compressed files. These files contain a header section and multiple value rows and are ideal for parallel processing. In the data acquisition phase, original genomics data is first transferred into the storage server (a cloud-based distributed storage system). The content of these files is first extracted and then fed to the extract, transform and load (ETL) module. The ETL component reads input genomics data files, applies a parallel, in-memory approach for transforming the data and provides the unstructured genomic data with a defined structure. After this stage, we can perform parallel in-memory processing operations on this data and can query without extracting the data from large unstructured genomic files each time. The implementation details of the genomics ETL module are further explained in Sect. 3.4.

Data Storage
Genomics data is unstructured (due to its variety of formats), non-relational, immutable, increasingly large in volume, high velocity and distributed in different geographic regions across the globe. This unstructured genomic data is transformed into a structured format in the data acquisition phase and stored in a NoSQL database.

This approach enables clinical/pharma research to query genomics data like any other data. Furthermore, the write-once approach also eliminates the need to read large unstructured genomic source files with each genomic data processing task and results in improving performance over the existing approaches that read genomic data from source for each analytic operation/cycle [29,31,32,37,50,52].

This approach to storing genomic data also provides a scalable fault-tolerant data storage and efficient data retrieval mechanism. The fault tolerance is provided by replicated storage. In the case of one region failure, the replicated data can easily be accessed and retrieved. Furthermore, it also provides scalable data distribution so that there is no need to move data from one geographic region to another region and thus avoids expensive data transfer (further explained in Sect. 3.2).

Crucial to data storage is a plug-in-play NoSQL datastore which will provide stateful storage for the transformed data obtained from ETL. The plug-in-play nature of this component will enable the utilization of different types of datastores for genomics, clinical/pharma, biomedical imaging from clinical and laboratory instrumentation, electronic medical records, wearables data and historical medical records.

Genomic Data Analyser

Genomic data can be analysed for knowledge discovery such as determining the role of genes in causing or preventing a disease, analysing patient's genetic making for personalised medicine, genome variation analysis, genotype clustering, gene expression microarrays, chromosome variations or gene linkage analysis.

In our framework, the genomic data analyser sits at the core and can perform any of the genomic data analysis tasks after data is successfully transformed into a NoSQL data store. It integrates machine learning libraries for analysing the genomic data. A typical end-to-end genomic data analysis pipeline is depicted in Fig. 3. Crucial to our approach, the analysis pipeline runs on a computing cluster with two key components, namely, fault-tolerant data parallelism and distributed in-memory processing. The ability to process and keep intermediate results in memory enables us to address I/O bottleneck. Another benefit of our approach is the reduced pre-processing time due to our data acquisition and storage approach.

The required genomic data is queried from a NoSQL datastore and ready for the analysis phase. The analysis phase applies the particular analysis algorithm(s) on the pre-processed loaded data. The genomics data analyser uses a fault-tolerant parallel in-memory computing cluster that can be used for batch and continuous near-real-time streaming jobs. The data acquisition is integrated as a module within the framework and are not limited to a specific type of analysis or a particular algorithm and can implement any algorithm for genomic data analysis. The plugin tools and library services of the genomic data analyser are explained in Fig. 2. The different algorithm categories that can be applied for genomic and clinical data analysis are:

- **Classification and Regression:** Linear models (logistic regression, linear regression and SVMs), Naive Bayes, Decision Trees, Gradient-Boosted and Random Forest Trees, Isotonic regression, Deep learning

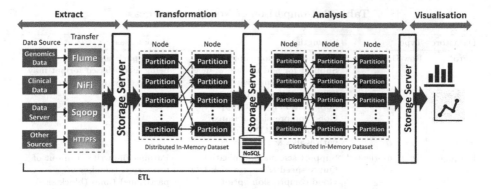

Fig. 3. Genomic data analytic pipeline

- **Collaborative Filtering:** Alternating Least Squares (ALS)
- **Decomposition:** Singular Value Decomposition (SVD), Principal Component Analysis (PCA)
- **Clustering:** K-means, Power Iteration Clustering, Latent Dirichlet Allocation, Gaussian Mixture

Examples of genomics data analysis using classification or regression algorithms include SVM for identification of functional RNA genes [6], Random forests to predict phenotypic effects of nucleotide polymorphisms [24]. Clustering algorithm K-means has been utilised for population-scale clustering of genotype information [4]. Deep learning has been utilised on variant calling tasks [44] and image to genetic diagnosis [20].

The list above provides some example categories of machine learning analytics that can be utilised with genomics data. Our framework is not limited to these algorithms and other algorithms can easily be integrated for analysing genomic and clinical data. The ability to access a varied list of supervised and unsupervised of algorithms increases the scope of analysis for genomic data.

Data Visualisation

Data visualisation is a web-based, language-independent component for data ingestion, exploration and visualisation. It is connected to both the analysis pipeline and the data storage server. We look at the two possible platforms for this component: Jupyter Notebook and Apache Zeppelin. Both can act as an interpreter for connecting to a variety of programming languages for backend data connectivity and visualisation. Jupyter Notebooks supports PySpark. Zeppelin is a Spark-based tool and naturally integrate with Apache Spark. Apache Zeppelin has the added advantages of integrating well with other big data tools such as Hadoop, HBase and Cassandra and provides support for PostgreSQL and Elastic Search.

A common weakness of the existing approaches (refer Sect. 2) is the lack of visualisation platforms. Even with genomics data analysis, little to no value

Table 3. Comparison of NoSQL datastores

Datastore	Type	Strengths	Weaknesses
Avro	Row-Based	Schema evolution supported, Support for add update & delete fields, Good language interoperability, Data Serialization & split support, Good compression options, Suitable for JSON, Compatible with Kafka & Druid	Poor Spark support, Avro being a row-based format may lead to slow query speed
Parquet	Column-oriented	Support schema optimisation, Query speed & reduce disk I/O, Good compression options, Support nested types in columns, Very good split support, Compatible with Spark, Arrow, Drill & Impala	Parquet file writes run out of memory if (number of partitions) times (block size) exceeds available memory
ORC	Column-oriented	Nested Data Types, Supports indexing, ACID transaction guarantees, Handle streaming data, Caching on the client side, Excellent split support, Compatible with Hive & Presto	Schema is partially-embedded in data
HBase	Column-oriented	Highly distributed, Fault tolerance, Schema-less, Fail over support & load sharing, SQL query API through Hive, Row-level atomicity	Single point of failure because of master-slave architecture, No Transaction support

can be obtained from unreadable information. By coupling analytic and storage components with a visualisation platform, the large volumes of data can be made comprehensible to end-user for valuable insights.

3.2   Data Centre Middleware

The data centre middleware layer enables a distributed, scalable and fault-tolerance for storage, management and efficient analysis of genomics data. This layer addresses the challenges of high processing time and the need for scalable data storage. It is capable of handling both structured and unstructured genomics data which make up the full range of common genomics data types. The services layer is built on top of a middleware that enables a distributed, scalable and fault-tolerance computing and storage for the management and analysis of genomics data. This layer consists of a storage server and computing cluster which provide distributed resource management and storage and distributed in-memory computing framework (Fig. 2).

Table 4. Data store compression codecs

Features	Snappy	Deflate	GZip	ZLib
Splittability Support	Avro, Paraquet, ORC	Avro	Paraquet	ORC
Tool Support	Avro, Paraquet, ORC, HBase	Avro	Paraquet, HBase	ORC
Design Emphasis	High compression Speed	Compression	Compression	Compression
Lossless compression	Lossless	Lossless	Lossless	Lossless

Storage Server

Storage server forms a key component of our framework. Unlike most traditional genomic data processing pipelines, we take a write-once and read-many approach to genomic data for ensuring its persistence. Genomic data is unstructured and massive in size, requiring a storage schema that caters for the volume, velocity and veracity of the generated data. The unstructured genomic data is read once from the input sources and written into distributed structured compact datasets. This approach enables the analysis pipeline to access data at a fraction of the time it would take read from the original datasets.

Genomics and clinical data are commonly available in unstructured formats. There is a diverse selection of NoSQL datastore available for different genomics and clinical data. Data are available in different levels of granularity ranging from key-value pairs, document-oriented, column/row format and even in graph database format. In key-value format, a byte array containing a value is addressed to a unique key. With this approach, scalability is emphasized over consistency. Document stores are schemaless, typically storing data in formats such as JSON and XML and have key-value pairs within a document. Document-oriented stores should be used to take advantage of the schemaless model. Column NoSQL stores are hybrid column/row and employ a massively distributed architecture. This approach enables massive scalability with improved read/write performance on large datasets. Graph databases emphasis the efficient management of strongly linked data, where operations on relationships can be made more efficient. The operations that are normally expensive, like recursive joins, can be replaced by traversals. Columnar NoSQL data stores are the preferred option for storing genomic data and provide a solution for scalability and query requirements.

Our storage server supports plug-in-play NoSQL datastores and is not limited to one particular type of NoSQL datastore. We evaluated Apache HBase (column-oriented, key-value store), Parquet (columnar storage), ORC (columnar storage), Avro (row-based format). All three Hadoop storage formats (Parquet, ORC and Avro) can provide significant space utilization savings in the form of compression, leading to reduced I/O in the analytic pipeline (Table 3).

Our storage server also allows the integration of relational databases to enable queries for extracting indexed data and perform further queries/operations and supports distributed datastores that meet the demands of high scalability and fault tolerance in the management and analysis of massive amounts of genomics/clinical data. The data is stored in a versioned format and any incremental changes in data can easily be tracked. This is very useful for time series data and can easily handle incremental changes in genomics data. It also provides the ability to make small reads, concurrent read/write and the incremental updates.

Splittability is an important feature for any datastore for computing frameworks with data parallelism as it affects the ability to partition read data. Having non-splittable chunks of data on a node prevents us from partitioning this data across a computing cluster, preventing efficient utilization of our executor resources.

Splittability of each of the compression codec available with the shortlisted storage solutions is summarised in Table 4. Snappy, Deflate and GZIP do not support splittability when used to compress plain text files but support splittability in a container format such as Hadoop file formats. ZLIB is only codec that splittable with plain text files. Splittability support on a distributed storage solution such as Hadoop is referred to as block-level compression. This is internal to the file format where individual blocks of data are compressed, allowing splittability with generally non-supportive codecs. Typical genomic analytic jobs on Spark Hadoop cluster are IO-bound, not CPU bound, so light and splittable compression codec will lead to improved run time performance.

Computing Cluster

Computing cluster is the foundation for data acquisition and genomics data analyser. It's built on top of high performance computing cluster which provides access to a collection of many nodes which are connected via a fast interconnected network file system and provides access processor cores and RAM.

To meet the challenges of genomics data outlined in Sect. 2, the computing cluster provides horizontal scalability and fault-tolerance to address the challenges of volume and velocity of genomics data and in-memory computing framework to reduce the disk I/O during analysis.

3.3   Data Centre Infrastructure Service

We have a cloud-based computing and storage cluster form our infrastructure layer. A cloud based approach gives us access to a large amount of on-demand resources, enabling us to increase the size of computing and storage cluster easily. Cloud computing is much more scalable and cost-efficient solution. A core enabler of cloud computing, virtualization technology facilitates maximum utilization of hardware and investment [5], allowing us to access massive resources at affordable prices. Advances in the robustness of distributed computing and parallelised programming also adds fault-tolerance to our cloud-based solution [56].

The infrastructure layer is composed of high-performance computing (HPC) cluster, scalable block & object storage cluster and high throughput network. A scalable object and block storage underpin the storage server in the middleware, enabling a highly distributed and fault tolerant storage service that's able to efficiently deal with the massive amounts of unstructured genomic and structured

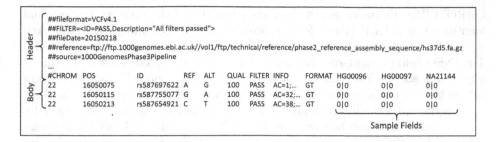

Fig. 4. Sample VCF File structure

clinical data. HPC cluster provides sufficient memory and core resources for the utilization of an in-memory and highly fault-tolerant distributed analytic pipeline.

3.4   Genomics Variant Analysis: A Case Study

We evaluated our framework with a genomic variant analysis case study. Genomic variant analysis detects genomics mutations related to a particular disorder in a human genome. This provides the quintessential challenges associated with genomics data and will test the in-memory, distributed and fault-tolerance of the proposed framework. Although this case study focuses on VCF file format, the framework can be used to handle all different file formats discussed in Sect. 2.1 and can integrate with existing clinical/pharma system. Genomic variant data is stored in VCF file format.

In this case study, we cluster individuals using the genotype information in VCF files. We implement two approaches, clustering with NoSQL (abbreviated as CwNoSQL) and clustering with VCF (abbreviated as CwVCF). Both these approaches involve parsing VCF data and either first writing the data to a NoSQL storage and then having the analyser access the variant data from NoSQL storage (clustering with NoSQL) or passing it straight to analyser (clustering with VCF), before performing the clustering operation. In the remainder of this section, we explain the sequence of the steps taken in implementing the case study for evaluating our framework.

Genotype clustering of variant data in VCF files involves the following steps: 1) acquiring genomics data and executing ETL module in data acquisition 2) pre-processing transformed data for clustering algorithm in VCF Data Pre-Processing and 3) genotype clustering using a clustering algorithm. These steps are explained further in the following sections.

Data Acquisition. Input VCF files are initially loaded into the genomics storage server. The genomics ETL module reads each VCF file as a text file and processes all the files row-by-row while skipping the header lines starting with "##" (refer Fig. 4). Individual field values are obtained by tab splitting a variant line from input files. The first eight columns in VCF file format contain the

CHROM (the chromosome), POS (starting position of the variant), ID (external identifier tags), REF (the reference position in the genome), ALT (the variant itself), QUAL (quality metrics), FILTER (quality control check indicators). The INFO field (variant-specific annotation information filed) is used by genome sequencing annotation tools to store vectors of annotations. The eighth column acts as the key to values from sample-specific data. All individual sample-specific data begins from the ninth column. There is no limit to the number of samples or annotations that can be contained in a VCF file. Each row in VCF has values attributed to the corresponding column in a VCF file. Sample columns depend on cohort size and are treated as individual columns in NoSQL datastore. This will enable fast lookup during pre-processing.

VCF Data Pre-processing. Following the data acquisition step, VCF data undergoes pre-processing. This step involves a number of transformations on the data in order to obtain the feature vectors required for executing genotype clustering. These transformations are based on the genotype pre-processing transformation steps [4]. The pre-processing stage for CwNoSQL and CwVCF differ with the latter having variant data read from VCF file directly while CwNoSQL will have the transformed data (refer Sect. 3.1) in NoSQL read. For CwVCF variant data is read line-by-line from the source and stored in arrays of to Resilient Distributed Datasets (RDDs). The RDDs allow creating an immutable distributed collection of objects across cluster providing fault-tolerance parallelized data. Tuples of values and their corresponding sample headers Table 5 are created using Spark's 'zip' function. This produces an array of elements containing key-value pairs (KVP) of sample and its header. KVPs that represent alleles are kept while other data are removed. The custom function converts KVP allele from strings in the form of '0—1' to doubles representing the hamming distance to the reference where a 0 is a no variant, 1 is heterozygous variant and a 2 is a homozygous variant. KVP with a value of 0, which represent a no variant data point, are removed as the data will be converted to sparse vectors in later steps. To further reduce the size of data we remove further data points that will contribute little end task. This can optionally include removing variants present in only one individual. This is done by removing arrays with a length of one as the length of this array represents the number of individuals that have the variant. To obtain arrays of alleles for each individual, we firstly zip array from the previous step with an ID to obtain ordered arrays of alleles. Spark's 'flatMap' is used to flatten arrays into KVPs. Key is the individual IDs while the value is the tuple of variant ID and variant information. The final step is to group these KVPs using individual IDs obtaining sparse vectors.

Genotype Clustering. For genotype clustering, we utilised Spark's MLlib which provides a variety of algorithms (refer Sect. 3.1) including K-means clustering algorithm. K-means model is fitted on all the features obtained in Sect. 3.4 using $k = 4$ centroids and $t = 100$ iterations. K-means is an unsupervised

Table 5. VCF file header explained

Header	Description
CHROM	The chromosome
#_Samples	Number of samples
Samples	Sample name
POS	Reference position from position 1 of the 1st base
ID	Semicolon separated list of unique identifiers
Ref	Reference base represented as A, C, G, T, N
QUAL	Phred scale assertion quality score for in ALT
ALT	Comma separated list of alternate non-reference alleles from a sample
AC	Allele count in genotype, for each ALT allele, in the same order as listed
AF	Allele frequency for each ALT allele
AN	Total number of alleles in called genotype

clustering algorithm and starts by randomly defining k centroids and proceeds to iteratively execute the following two tasks:

1. Using Euclidean distance, each data point is allocated to the nearest centroid
2. Minimizing the mean value of cluster centroids with cluster data points

The re-grouping process continues until no new improvements can be performed. K-means algorithm is relatively efficient with the complexity of $O(tkn)$, where n is numbers of objects.

4   Empirical Performance Evaluation

In this section, we first present details of our experimental setup followed by the empirical performance evaluation. This set of results focus on evaluating and benchmarking storage and analytic components of the framework. The experiments focus on evaluating the performance of our framework on extract, transform and load, compression, datastore read, random data lookup, task parallelism, data splitting, partition support and genotype clustering. We also

Table 6. Data-write comparison with data stores at varying storage intervals

Datastores	Write time (seconds)				
	20%	40%	60%	80%	100%
HBase	521.4	1028.1	2309.9	3393.2	4737.7
Parquet	113.2	221.3	260.6	367.2	456.6
ORC	125.5	239.4	330.5	384.6	480.6
Avro	107.4	188.9	279.6	365.6	453.2
Text (CSV)	246.7	492.1	716.0	883.6	1178.6

Table 7. Data read comparison

Datastore	Time (mm:ss)	
	Chromosome 22	Chromosome 1
HBase	00:27	06:52
Parquet	00:17	00:23
ORC	00:18	00:35
Avro	00:28	00:48
VCF	01:14	10:57
CSV	00:34	05:01

compare our framework with VariantSpark on genotype clustering task and conclude this section with recommendations on specific components for the proposed framework.

4.1 Experimental Setup

This section explains the implementation and experimental details for evaluating the presented framework. The results evaluate all the stages of our framework from reading input data files, transforming the data into a structured format, transforming and storing the data in the storage server, implementing analytics algorithm(s) on the pre-processed data and interactively representing the results.

Evaluation Parameters. The performance of the storage server for transforming and managing genomics data is measured with the following aspects: 1) extract, transform and load performance (a key omnipresent task for genomics data volume and velocity), 2) read performance which delves into datastore performance in querying, 3) random data lookup which provides a snapshot of datastore lookup latency, 4) data splitting and partition support probing datastore scalability and fault-tolerance.

To evaluate the framework's analytic components we evaluated the following tasks: 1) ETL that tested compute engine's scalability and fault-tolerance 2) data sorting (a key test for computing component as it involves many of the transformations that are present in analytic tasks, testing parallelism, in-memory and I/O performance), 3) genotype clustering as it provides the most complete test of the analytic and storage components of the proposed framework. Scalability and performance of the framework are evaluated by increasing the number of computing cores from 8–80.

Dataset. The result reported in this paper are obtained using phase 1[4] and phase 3[5] datasets from the 1000 genome project. The total size of Phase 1 and

---

[4] http://ftp.1000genomes.ebi.ac.uk/vol1/ftp/release/20110521/.
[5] http://ftp.1000genomes.ebi.ac.uk/vol1/ftp/release/20130502/.

Phase 3 data, in their raw uncompressed VCF format, are approximately 161 GB and 770 GB. It contains 84.4 million variants [9] for 1092 and 2504 individuals respectively, these samples are from 26 populations and belong to five super populations as African (AFR), American (AMR), East Asian (EAS), European (EUR) and South Asian (SAS) [9].

Compute Cloud. The experiments were conducted on a compute cluster consisting of 88 CPUs and 176 GB RAM. These cluster resources are obtained from the University of Derby compute cloud. This cluster is composed of servers equipped with Intel Xeon processors where each processor consists of 6 physical cores and is running a 64-bit instruction set. The compute cluster is running OpenStack Icehouse with Ubuntu 14.0.4 LTS. All nodes are interconnected via a Network File System (NFS) using a dedicated 10 GB/s Ethernet link and switches to share data between the servers.

It is important to note here that our framework is robust and can run on any computing and storage cluster configuration. The computing resources used for these experiments do not represent a minimum or maximum requirement for using our framework. These resources can be extended or reduced as required for a particular experiment.

ETL Performance. We evaluate extract transform and load performance of the framework on multiple datastores in 20% dataset load intervals. We looked at five potential datastore solutions: Apache HBase, Apache Parquet, Apache ORC and Apache Avro. These are compared to the original VCF file format and standard CSV format. We compared the performance of these datastores by bulk loading phase 3 chromosome-1 uncompressed VCF dataset

Results in Table 6 demonstrate the ETL performance of Avro, Parquet and ORC which were approximately 10 times quicker than HBase and less than half of the time for CSV, with Avro and Parquet having 27 and 24 s over ORC. Avro is a row-based datastore that provides it with an edge over other datastores on write operations as it is quicker to append records. For the Hadoop storage formats, we can parse the VCF file for the variants to a dataframe, a Spark

Table 8. Datastore compression comparison

Datastore	Size on disk (GB)
HBase	70.4
Parquet	5.9
ORC	3.6
Avro	4.9
CSV	66.7
VCF	65.7

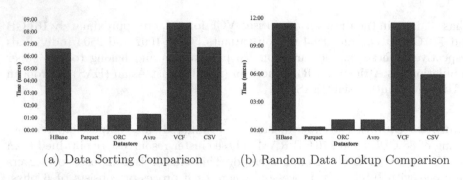

(a) Data Sorting Comparison   (b) Random Data Lookup Comparison

Fig. 5. Data sorting and random data lookup comparison

distributed data structure. This is then written to the storage format with a function RDD write function. HBase requires additionally mapping each variant to an HBase put object which increases runtime. Spark's native support for writing dataframes to Hadoop storage formats greatly simplifies the process of writing data.

Data Compression Performance. We investigated the compression performance for each of the datastores. Apache Parquet, ORC and Avro formats offer a variety of data compression options. Avro supports Snappy and Deflate compression, while Parquet supports Snappy and GZIP and ORC supports Snappy and ZLIB. ORC, Avro and Parquet consumed only 5.5%, 7.5% and 9% of the original VCF file disk space respectively while HBase and CSV exceeded 100% of original disk space (refer Table 8). ORC outperformed Parquet and Avro by 39% and 27% respectively. ORC outperforms both because it's default compression codec ZLib, provides better compression ratios compared to Parquet's Snappy and Avro's Deflate.

Datastore Read Performance. We compared the time for reading the data and counting the number of records in each dataset for all the NoSQL datastores (Apache Parquet, Apache ORC, Apache Avro, Apache HBase) and VCF file format. Table 7 displays read performance for the aforementioned datastores. Parquet and ORC provided the best time followed by HBase and Avro with VCF file read last. All three Hadoop data formats benefit from compression but it's the columnar data storage that gives Parquet and ORC the edge in data retrieval. It enables efficient lookup of a subset of columns, reducing I/O. Parquet and ORC are optimized for write-once-read-many strategy.

Data Sorting Performance. This task is one of the most important operations that requires moving data across nodes in the cluster. Sorting a large volume of data requires moving all the data across the network, testing the scalability and fault- tolerance of the computing cluster and disk I/O. In an in-memory computing framework, shuffle operation predicates nearly all distributed

Table 9. Task parallelism with varying data partitions

Number of data partitions	Number of allocated CPUs		
	36	45	54
9	12:38	12:53	12:14
18	12:31	13:05	12:27
36	12:11	11:51	11:53
72	10:50	10:24	10:13
144	09:24	09:13	08:57
288	07:43	07:34	07:34
576	08:36	08:34	08:26
1152	09:02	08:52	08:33

processing workloads as it tests the parallelism, fault-tolerance and in-memory performance of the computing framework. Unlike MapReduce, Spark's sorting involves no reduction of data along the pipeline, making it more challenging.

We found Parquet had a 3 and 8 s over ORC and Avro (refer Fig. 5a). VCF and CSV lagged far behind the Hadoop datastores and HBase fared even worse. Although both VCF and CSV are row-based text files, Spark's built-in support for reading CSV data provided CSV with performance boost over VCF. These results are in line with datastore read comparison and show the edge columnar datastores have over row-based Avro, VCF and CSV when querying data.

Random Data Lookup Performance. We investigated random data lookup for each of the datastores. Using a compound key, in this case, the 'pos' identifier in the variant data (refer Fig. 4), we retrieved a variant record in phase 3 chromosome 1 dataset (65.7 GB). Figure 5b depicts results with Parquet ahead of ORC and Avro by 44 and 45 s, followed by a big gap with HBase, VCF and CSV. For row-based Avro and CSV, individual records can only be looked up using a brute force scan of the read data. Columnar datastores like Parquet and ORC, on the other hand, are optimized for data lookup, taking advantage of predicate and projection pushdown which translate to the efficient manner in which columnar datastores can filter and omit unnecessary data from table scans or queries in large volumes of stored data.

Task Parallelism Performance. With a distributed computing and storage server, resources are not utilised efficiently when the level of parallelism is not sufficient for each job. The input data is divided into partitions for parallel in-memory processing. These partitions are granular chunks of data that cannot span across multiple nodes on the cluster. Apache Spark's default setting is configured to create one partition for each HDFS block from input files which is set to 64 MB by default. However, the block size is not fixed and can vary

22   T. Abdullah and A. Ahmet

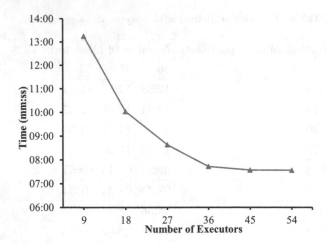

Fig. 6. Data load runtime with varying executors

depending on the cluster size. Spark reads data into an RDD from the nodes
that are close to it for minimizing network traffic and disk IO.

Creating too few partitions will lead to less concurrency and data skewing
that will result in performance bottlenecks. Similarly, creating too many partitions can also to lead to increased runtime due to task scheduling overhead.

To evaluate task parallelism, we executed experiments with a varying number
of partitions and executor allocations for our input genomics data using VCF
data ETL process. We read phase-3 chromosome-1 VCF dataset split into 130
individual VCF files (65.7 GB), transform and write the data to parquet format.
In these experiments, we varied the number of partitions between 9 and 1152
and the number of executors are varied between 9 to 54.

We found all executor allocations performed poorly with 9, 18 and 36 partitions but performance doubled from 36 to 72 and 72 to 144 (refer Table 9).
The best performance was observed with 288 partitions and followed by 576
and 1152. These results show the number of partitions and core allocation
greatly affects the parallelism of a task and ultimately the runtime performance.

Table 10. Datastore compression codecs

Compression codec	Splittability support	Tool support	Design emphasis
Snappy	Avro, Paraquet, ORC	Avro, Paraquet, ORC, HBase	Compression speed
Deflate	Avro	Avro	Compression ratio
GZip	No	Paraquet, HBase	Compression ratio
ZLib	ORC	ORC	Compression ratio

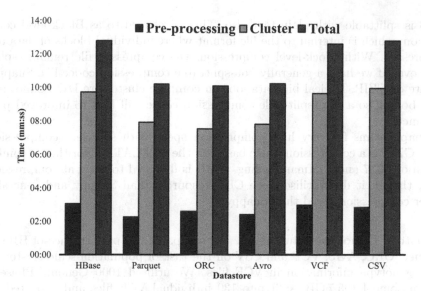

Fig. 7. Genotype clustering comparison with data storage

Having too few partitions will lead to less concurrency and data skewing which will result in performance bottlenecks. Having too many partitions can also to lead to increased runtime due to task scheduling overhead.

Spark is optimized to create partitions from data stored in nearby nodes. The partitions are also located on worker nodes that execute tasks. The optimal executor memory allocation strategy for spark jobs was 2 GB memory per executor. It was also observed that this memory footprint increases with increasing genome coverage depth. Picking the optimal number of partitions also depend on cluster size and resource allocation.

We also evaluated the scalability of the proposed framework using a varying number of executors using ETL operation. We observed runtime dropped approximately 24% with 9–18 executors, 14% with 18–27 and 11% with 27–36 executors before starting to plateau with 36–54 executors (refer Fig. 6). The ETL operation involves reading a large number of files distributed over storage server and executing numerous Spark transformations that test key aspects of computing server parallelism.

Data Splitting and Partition Support. Splittability is important for computing frameworks with data parallelism as it affects our ability to partition our data. In essence, if a datastore does not support splittable chunks of data on a node, this will prevent us from partitioning this data using and limit the efficient utilization resource allocated to a task.

Table 10 displays the splittability for each of the compression codecs. Snappy, Deflate and GZIP are not splittable when compressing plain text files but support splittability if in a container format such as Avro, Parquet or ORC, while

ZLIB is splittable with plain text also. This is referred to as Block-level compression, which is internal to the file format where individual blocks of data are compressed. With Block-level compression, the compressed file remains splittable even if we have a generally non-splittable compression codec like Snappy, Deflate or GZIP. Typical big data jobs on compute cluster are I/O bound, not CPU bound, so a fast, splittable compression codec will lead to improved performance.

Snappy aims for very high compression speed with sufficient compression ratio. GZIP is a compression codec based on the DEFLATE algorithm, a combination of LZ77 and Huffman Coding. GZIP is designed to excel at compression ratio, though it does utilise more CPU resources than Snappy and generally slower compression speed than Snappy.

Datastore Genotype Clustering. We compared the performance of HBase, Parquet, ORC, Avro, VCF and CSV on the task of population-scale clustering using genotype information in VCF files. We utilised 1000 Genome Phase 3 Chromosome 1 (65.7 GB) split into 130 individual VCF files and allocated 36 cores and 72 GB of memory (refer Fig. 7). We found ORC, Parquet and Avro performed with similar times. Though ORC recorded the lowest runtime, all four displayed very close performance, with 11 s between ORC and Avro. These were followed by HBase, VCF and CSV, also displaying similar times but significantly higher than ORC, Parquet and Avro.

These results demonstrate the edge ORC, Parquet and Avro in analytic tasks which process a subset of columns or rows. For genotype clustering, instead of parsing VCF or CSV files row by row and executing transformations to obtain columns of genotype sample data (refer Sect. 3.4), we can query pre-processed columns in ORC, Parquet and Avro and skip key transformations. This advantage combined with efficient compression means NoSQL datastore can potentially read 18x less data with substantially less expensive processing required for an analytic task such as genotype clustering.

Comparison with VariantSpark. Unlike the VariantSpark implementation, our genotype clustering approach (clustering-Parquet) does not perform direct transformations on VCF files, rather, we parse VCF data and store in the genomics storage server once. Our approach, clustering with Parquet was markedly faster than VariantSpark (refer Fig. 8). This approach resulted in performance gains in genotype pre-processing.

Genomic data in VCF files are unstructured, VariantSpark parses this data line by line for every analytic task. Whereas, in out framework, we parse the VCF data once, giving it structure and optimizing for genotype query tasks, and store it into our genomics storage server. This means that all data reads will be quicker with our genomics datastore as we can efficiently lookup whole sample columns from datastore and avoid reading all the data, row by row, and executing costly transformations described in Sect. 3.4. These costly transformations are executed once during data acquisition and stored in datastore, making them

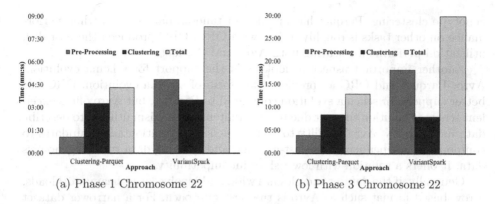

(a) Phase 1 Chromosome 22   (b) Phase 3 Chromosome 22

Fig. 8. Genotype clustering comparison with VariantSpark

available for future analysis. Another key benefit of our approach is that the analysis results are stored in the genomics storage server and are not discarded as in VariantSpark. Our approach yields long-term benefits, as we don't need to parse VCF files for each analytic task on the genomics dataset. Another advantage of storing the parsed VCF data into our genomics storage server is massive scalability and a drastic reduction in disk utilization as a result of efficient compression. Crucially, a columnar NoSQL datastore approach enables us to scale horizontally on commodity hardware and partitioning across multiple servers. With growing data and increasing storage demand, we can add nodes to the existing server.

Results Discussion and Recommendations. In this section, we provide a discussion of our findings on the case study findings and provide key recommendations for the proposed framework architecture. In our case study, we utilised Apache Spark for computing framework and Apache Hadoop and plug-and-play NoSQL datastore for storage server. Spark offers a distributed, fault-tolerant, in-memory computing framework to meet the challenges of genomics data. Spark integrates naturally with Apache Hadoop and utilises Hadoop's YARN in cluster mode. We utilised HDFS for the backbone of the storage server and a foundation to the genomic datastore. Hadoop's YARN also served as a resource manager for Spark computing framework. The combination of Hadoop and Spark is a proven recipe for massively scalable and fault-tolerant computing and storage.

Selecting the right datastore for a storage server should be based on the following considerations: write performance, partial/full read performance, disk utilization, splittability, data compression performance and tool support. Datastores that are not splittable should be avoided as this will prevent the utilization of compression codecs.

We recommend choosing a columnar datastore such as Parquet or ORC over row-based formats Avro or CSV or any text-based dataset format like VCF. Parquet and ORC have demonstrated superior performance over other datastores in datastore read, data sorting, random data lookup, data splittability and genotype clustering. Parquet has an edge in random data lookup while performance on other tasks is roughly in line with ORC. ORC produced the best disk utilization performance beating both Avro and Parquet.

Another important aspect to look at is the support for schema evolution. Avro, Parquet and ORC all provide some form of schema evolution. ORC has better support for schema evolution compared to Parquet, but Avro offers excellent schema evolution support due to the inherent mechanism utilised to describe data using JSON. Avro's ability to manage schema evolution allows updates to components independently, for fast-growing and evolving data such as genomics data, it offers a solution with low risk of incompatibility.

Generally, if the dataset samples are wide and involve write-heavy workloads, a row-based format such as Avro is the best approach. For a narrower dataset with ready-heavy analytic workloads, a columnar approach is recommended. Sometimes Hadoop users may be committed to a column-based format but as they start getting into a project, the I/O pattern starts to shift towards a more write-heavy presence. In that case, it might be better to switch over to row-based storage (Avro) but add indexes that provide better read performance.

Different Hadoop tools also have different affinities for Avro, ORC and Parquet. ORC is commonly used with Apache Hive (managed by Hortonworks). Therefore, ORC data tends to congregate in companies that run the Hortonworks Data Platform (HDP). Presto is also affiliated with ORC files. Similarly, Parquet is commonly used with Impala, and since Impala is a Cloudera project, it's commonly found in companies that use Cloudera's Distribution of Hadoop (CDH). Parquet is also used in Apache Drill, which is MapR's favored SQL-on-Hadoop solution; Arrow, the file-format championed by Dremio; and Apache Spark, everybody's favourite big data engine that does a little of everything.

Avro, by comparison, is the file format often found in Apache Kafka clusters, according to Nexla. Avro is also the favoured big data file format used by Druid (the high-performance big data storage and compute platform). What file format you use to for data storage in your big data solution is important, but it's just one consideration among many.

Genomics datasets have high disk utilization due to the vast volume of data and data replication of distributed storage platform. To add further scalability to existing storage tools we recommend efficient compression, enabling a significant reduction in disk utilization by up to 18 times for VCF datasets. The reduced disk utilization addresses another key genomics data bottleneck of disk I/O in the analytic pipeline. This leads to further improvement in runtime and data lookup latency. ORC outperformed both Avro and Parquet on compression and is the clear choice in this area. Based on our experiments, columnar datastores such as Apache ORC and Apache Parquet demonstrated an excellent balance between efficient data read, random data lookup, data ingestion, compression and scalable data analytics. It was also observed that datastores with indexing support made data retrieval easier. Table 8 summarises the disk utilisation for each data store.

5   Conclusions and Future Work

In this paper, we present a framework for efficiently tackling the challenges of genomics data on a Big data infrastructure. Unstructured genomic data is read from sources once, transformed, given structure and stored in optimized datastore. Medical and clinical data exist in a multitude of formats and repositories. Storage solutions that can accommodate heterogeneous genomic and clinical data will be key in developing analytic capabilities for producing a personalized treatment. Utilizing a NoSQL datastore on top of a highly proven distributed storage management framework ensures our approach can provide massive scalability with high fault tolerance.

Genotype clustering is presented as a case study for evaluating our storage server and compute framework. The results demonstrate superior performance over existing approaches which utilise architectures that do not effectively address storage and computing challenges of genomics data.

We evaluated datastores for storage server and found columnar storage formats such as Parquet and ORC provided the best read performance over row-based format Avro and column-oriented HBase. Flexible compression options from ORC, Avro and Parquet can provide significant disk utilization improvement. Spark was reading at least 10 times less data versus HBase and VCF file format. More efficient compression is important in reducing IO and resource utilization.

These results lead us to determine Parquet being the most suitable for scalable genomics data analytics. While Parquet and ORC are more computationally expensive in writing, Parquet provides the most promising data lookup latency.

In our future work, we plan to use state of the art deep learning algorithms with the proposed framework on a GPU cluster utilizing a variety of genomics, clinical and health data for healthcare prediction tasks. Recent advances have made deep learning one of the hottest technology trends [27] even surpassing human-level performance on certain classification and regression tasks. We believe deep learning will play an increasingly important role in the process of extracting value from genomics data.

Appendix A

In this section we provide detailed setup for the experiments conducted in genotype clustering case study. All experiments were conducted on Derby University IaaS platform. These experiments can be replicated on any cloud service or on-premise data centre architecture.

Resource Infrastructure. The cluster utilised for experiments consisted of 11 nodes with 8 CPU, 16 GB RAM and 200 GB Disk each providing a total of 88 processors, 176 GB RAM and 2.2 TB of disk. Infrastructure is interconnected with a Network File System (NFS) using a dedicated 10 GB/s ethernet link and switches to share data between the servers. In our experiments, we refer to allocated computing cores as executors.

Software Infrastructure. Each node ran Openstack Icehouse on Ubuntu 14.0.4 Server 64-bit. We utilised the following tools which populated the framework: Apache Hadoop 2.7.3, Apache Spark 2.1.1, Apache HBase 1.2.1, Apache Zeppelin 0.7.2, Java 1.8 and Scala 2.11 on every node. Each slave node ran yarn manager, HDFS data node and spark worker service for all experiments. Some of the experiments required HBase service. With a total of 11 nodes, we utilised 1 master and 10 slave node architecture (refer Fig. 9).

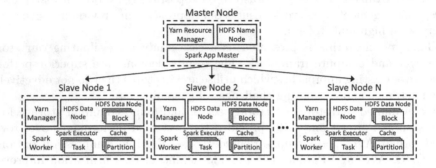

Fig. 9. Cluster architecture: implementation view

Dataset and Source Code. Experiment result were obtained using variant data available in VCF files from phase 1[6] and phase 3[7] datasets of the 1000 genome project. We are unable to release the full source code from this study as we are currently utilizing it in a related study and to develop it as a product.

References

1. Abdullah, T., Ahmet, A.: Genomics analyser: a big data framework for analysing genomics data. In: Proceedings of the Fourth IEEE/ACM International Conference on Big Data Computing, Applications and Technologies, pp. 189–197 (2017)
2. Bateman, A., Wood, M.: Cloud computing. Bioinformatics **25**(12), 1475 (2009)
3. Benson, D.A., Karsch-Mizrachi, I., Lipman, D.J., Ostell, J., Sayers, E.W.: GenBank. Nucl. Acids Res. **37**(Database), D26–D31 (2009)
4. Brien, A.R.O., Saunders, N.F.W., Guo, Y., Buske, F.A., Scott, R.J., Bauer, D.C.: VariantSpark: population scale clustering of genotype information. BMC Genomics **16**, 1–9 (2015)
5. Shaffer, C.: Next-generation sequencing outpaces expectations. Nat. Biotechnol. **25** (2007)
6. Carter, R.J., Dubchak, I., Holbrook, S.R.: A computational approach to identify genes for functional RNAs in genomic sequences. Nucl. Acids Res. **29**(19), 3928–3938 (2001)

---

[6] http://ftp.1000genomes.ebi.ac.uk/vol1/ftp/release/20110521/.
[7] http://ftp.1000genomes.ebi.ac.uk/vol1/ftp/release/20130502/.

7. Hayden, E.C.: Genome researchers raise alarm over big data. Nature (2015)
8. Chen, X., Jorgenson, E., Cheung, S.: New tools for functional genomic analysis. Drug Discov. Today **14**(15), 754–760 (2009)
9. The 1000 Genome Project Consortium: A global reference for human genetic variations. Nature **256**, 68–78 (2015)
10. Cook, C.E., Bergman, M.T., Cochrane, G., Apweiler, R., Birney, E.: The European bioinformatics institute in 2017: data coordination and integration. Nucl. Acids Res. **29**(19), 3928–3938 (2017)
11. Coonrod, E., Margraf, R., Russell, A., Voelkerding, K., Reese, M.: Clinical analysis of genome next-generation sequencing data using the Omicia platform. Expert. Rev. Mol. Diagn. **13**(6), 529–540 (2013)
12. Davies, K.: The 1,000 Dollar Genome - The Revolution in DNA Sequencing and the New Era of Personalized Medicine. Free Press (2010)
13. de Paula, R., Holanda, M., Gomes, L.S.A., Lifschitz, S., Walter, M.E.M.T.: Provenance in bioinformatics workflows. In: BMC Bioinformatics Workshops (2013)
14. Decap, D., Reumers, J., Herzeel, C., Costanza, P., Fostier, J.: Halvade: scalable sequence analysis with MapReduce. Bioinformatics **31**(15), 2482–2488 (2015)
15. Ding, L., Wendl, M., Koboldt, D., Mardis, E.: Analysis of next-generation genomic data in cancer: accomplishments and challenges. Hum. Mol. Genet. **19**(2), 188–196 (2010)
16. EMBL-EBI. EMBL-EBI annual scientific report 2013. Technical report, EMBL-European Bioinformatics Institute (2014)
17. Borozan, I., et al.: CaPSID: a bioinformatics platform for computational pathogen sequence identification in human genome and transcriptomes. BMC Bioinform. **13**, 1–11 (2012)
18. National Center for Biotechnology Information. File format guide, U.S. National Library of Medicine. https://www.ncbi.nlm.nih.gov/sra/docs/submitformats/
19. Guo, X., Meng, Y., Yu, N., Pan, Y.: Cloud computing for detecting high-order genome-wide epistatic interaction via dynamic clustering. BMC Bioinform. **15**(1), 102 (2014)
20. Gurovich,,Y., et al.: DeepGestalt-identifying rare genetic syndromes using deep learning. arXiv preprint arXiv:1801.07637 (2018)
21. Huang, H., Tata, S., Prill, R.J.: BlueSNP. R package for highly scalable genome-wide association studies using Hadoop clusters. Bioinformatics **29**(1), 135–136 (2013)
22. Huang, L., Kruger, J., Sczyrba, A.: Analyzing large scale genomic data on the cloud with Sparkhit. Bioinformatics **34**(9), 1457–1465 (2017)
23. Data — 1000 Genomes. IGSR: The International Genome Sample Resource. https://www.internationalgenome.org/data
24. Tian, J., Wu, N., Guo, X., Guo, J., Zhang, J., Fan, Y.: Predicting the phenotypic effects of non-synonymous single nucleotide polymorphisms based on support vector machines. BMC Bioinform. **8**, 450–546 (2007)
25. Jourdren, L., Bernard, M., Dillies, M.A.L., Crom, S.: Eoulsan. A cloud computing-based framework facilitating high throughput sequencing analyses. Bioinformatics **28**(11), 1542–1543 (2012)
26. Kelly, B.J., et al.: Churchill: an ultra-fast, deterministic, highly scalable and balanced parallelization strategy for the discovery of human genetic variation in clinical and population-scale genomics. Genome Biol. **16**(1), 6 (2015)
27. Klinger, J., Mateos-Garcia, J.C., Stathoulopoulos, K.: Deep learning, deep change? Mapping the development of the artificial intelligence general purpose technology. Mapp. Dev. Artif. Intell. Gen. Purp. Technol. (2018)

28. Kozanitis, C., Patterson, D.A.: GenAP: a distributed SQL interface for genomic data. BMC Bioinformat. **17**(63) (2016)
29. Langmead, B., Trapnell, C., Pop, M., Salzberg, S.L.: Ultrafast and memory-efficient alignment of short DNA sequences to the human genome. Genome Biol. **10**, R25 (2009). https://doi.org/10.1186/gb-2009-10-3-r25
30. Langmead, B., Schatz, M.C., Lin, J., Pop, M., Salzberg, S.L.: Searching for SNPs with cloud computing. Genome Biol. **10**(11), 134:1–134:10 (2009)
31. Langmead, B., Schatz, M.C., Lin, J., Pop, M., Salzberg, S.L.: Searching for SNPs with cloud computing. Genome Biol. **10**(11), R134 (2009)
32. Lu, W., Jackson, J., Barga, R.: AzureBlast: a case study of developing science applications on the cloud. In: 19th ACM International Symposium on High Performance Distributed Computing, pp. 413–420 (2010)
33. Mardis, E.R.: The impact of next-generation sequencing technology on genetics. Trends Genet. **24**(3), 133–141 (2008)
34. Massie, M., et al.: Adam: genomics formats and processing patterns for cloud scale computing. Technical report UCB/EECS-2013-207, EECS Department, University of California, Berkeley, December 2013
35. Mohammed, E.A., Far, B.H., Naugler, C.: Applications of the MapReduce programming framework to clinical big data analysis: current landscape and future trends. BioData Min. **7**(1), 1–23 (2014)
36. Wiewiorka, M.S., Messina, A., Pacholewska, A., Maffioletti, S., Gawrysiak, P., Okoniewski, M.J.: SparkSeq: fast, scalable and cloud-ready tool for the interactive genomic data analysis with nucleotide precision. Bioinformatics **15**(30), 2652–2653 (2014)
37. Nordberg, H., Bhatia, K., Wang, K., Wang, Z.: BioPig: a Hadoop-based analytic toolkit for large-scale sequence data. Bioinformatics **29**(23), 3014–3019 (2013)
38. Norrgard, K.: Genetic variation and disease: GWAS. Nat. Educ. **1**(1), 87(2008)
39. O'Connor, B.D., Merriman, B., Nelson, S.F.: SeqWare query engine: storing and searching sequence data in the cloud. BMC Bioinform. **11**(Suppl. 12), S2 (2010)
40. Oliveira, J.H., Holanda, M., Guimaraes, V., Hondo, F., Filho, W.: Data modeling for NoSQL based on document. In: Second Annual International Symposium on Information Management and Big Data, pp. 129–135 (2015)
41. Pinheiro, R., Holanda, M., Arujo, A., Walter, M.E.M.T., Lifschitz, S.: Automatic capture of provenance data in genome project workflows. In: IEEE International Conference on Bioinformatics and Biomedicine (BIBM), pp. 15–21 (2013)
42. Pinherio, R., Holanda, M., Araujo, A., Walter, M.E.M.t., Lifschitz., S.: Storing provenance data of genome project workflows using graph databases. In: IEEE International Conference on Bioinformatics and Biomedicine (BIBM), pp. 16–22 (2014)
43. Pireddu, L., Leo, S., Zanetti, G.: Seal: a distributed short read mapping and duplicate removal tool. Bioinformatics **27**(15), 2159–2160 (2011)
44. Poplin, R., et al.: A universal SNP and small-indel variant caller using deep neural networks. Nat. Biotechnol. **36**(10), 983–987 (2018)
45. 1000 Genomes Project. Data types and file formats
46. Zou, Q., Li, X.B., Jiang, W.R., Lin, Z.Y., Li, G.L., Chen, K.: Survey of MapReduce frame operation in bioinformatics. Brief. Bioinform. **15**, 637–647 (2014)
47. Qiu, J., et al.: Hybrid cloud and cluster computing paradigms for life science applications. BMC Bioinform. **11**(12), 1–6 (2010). BioMed Central
48. Quail, M.A., et al.: A tale of three next generation sequencing platforms: comparison of Ion Torrent, Pacific Biosciences and Illumina MiSeq sequencers. BMC Genomics **13**(1), 1–13 (2012). BioMed Central

49. Robinson, T., Killcoyne, S., Bressler, R., Boyle, J.: SAMQA: error classification and validation of high-throughput sequenced read data. BMC Genomics **12**, 419 (2011)

50. Schatz, M.C.: Cloudburst: highly sensitive read mapping with MapReduce. Bioinformatics **25**(11), 1363–1369 (2009)

51. Schoenherr, S., Forer, L., Weissensteiner, H., Specht, G., Kronenberg, F., Kloss-Brandstaetter, A.: Cloudgene: a graphical execution platform for MapReduce programs on private and public clouds. BMC Bioinform. **13**(1), 200 (2012)

52. Schumacher, A., et al.: SeqPig: simple and scalable scripting for large sequencing data sets in Hadoop. Bioinformatics **30**(1), 119–120 (2014)

53. Stein, L.D.: The case for cloud computing in genome informatics. Genome Biol. **11**(5), 207 (2010)

54. Stephens, Z.D., et al.: Big data: astronomical or genomical? PLoS Biol. **13**(7), e1002195 (2015)

55. Taylor, R.C.: An overview of the Hadoop/MapReduce/HBase framework and its current applications in bioinformatics. BMC Bioinform. **11**(S12), S1 (2010). Springer

56. Wong, K.-C., Zhang, Z.: SNPdryad: predicting deleterious nonsynonymous human SNPs using only orthologous protein sequences. Bioinformatics **30**(8), 1112–1119 (2014)

57. Yin, Z., Lan, H., Tan, G., Lu, M., Vasilakos, A., Liu, W.: Computing platforms for big biological data analytics: perspectives and challenges. Comput. Struct. Biotechnol. J. **15**, 403–411 (2017)

Dynamic Estimation and Grid Partitioning Approach for Multi-objective Optimization Problems in Medical Cloud Federations

Trung-Dung Le[1]([envelope]) [ID], Verena Kantere[2] [ID], and Laurent d'Orazio[1] [ID]

[1] Univ Rennes, 2 rue du Thabor - CS 46510 - 35065 Rennes CEDEX, Rennes, France
{trung-dung.le,laurent.dorazio}@irisa.fr
[2] University of Ottawa, 75 Laurier Ave E, Ottawa, ON K1N 6N5, Canada
vkantere@uOttawa.ca

Abstract. Data sharing is important in the medical domain. Sharing data allows large-scale analysis with many data sources to provide more accurate results. Cloud federations can leverage sharing medical data stored in different cloud platforms, such as Amazon, Microsoft, etc. The pay-as-you-go model in cloud federations raises important issues of Multi-Objective Optimization Problems (MOOP) related to users' preferences, such as response time, money, etc. However, optimizing a query in a cloud federation is complex with increasing the variety, especially due to a wide range of communications and pricing models. The variety of virtual machines configuration also leverages the high complexity in generating the space of candidate solutions. Indeed, in such a context, it is difficult to provide accurate estimations and optimal solutions to make relevant decisions. The first challenge is how to estimate accurate parameter values for MOOPs in a cloud federation consisting of different sites. To address the accurate estimation of parameter values problem, we present the Dynamic Regression Algorithm (DREAM). DREAM focuses on reducing the size of historical data while maintaining the estimation accuracy. The second challenge is how to find an approximate optimal solution in MOOPs using an efficient Multi-Objective Optimization algorithm. To address the problem of finding an approximate optimal solution, we present Non-dominated Sorting Genetic Algorithms based on Grid partitioning (NSGA-G) for MOOPs. The proposed algorithm is integrated into the Intelligent Resource Scheduler, a solution for heterogeneous databases, to solve MOOP in cloud federations. We validate our algorithms with experiments on a decision support benchmark.

Keywords: Cloud computing · Multiple Linear Regression · Cloud federations · Genetic algorithm · Non-dominated Sorting Genetic Algorithm

© Springer-Verlag GmbH Germany, part of Springer Nature 2020
A. Hameurlain and A M. Tjoa (Eds.): TLDKS XLVI, LNCS 12410, pp. 32–66, 2020.
https://doi.org/10.1007/978-3-662-62386-2_2

1   Introduction

Cloud federation is a paradigm of interconnecting the cloud environments of more than one service providers for multiple purposes of commercial, service quality, and user's requirements [35]. Besides of vendor lock-in and provider integration, a cloud federation has several types of heterogeneity and variability in the cloud environment, such as wide-range communications and pricing models.

Cloud federations can be seen as a major progress in cloud computing, in particular for the medical domain. Indeed, sharing medical data would improve healthcare. Federating resources makes it possible to access any information even on distributed hospital data on several sites. Besides, it enables to access larger volumes of data on more patients and thus provide finer statistics.

For example, patient A has just come back from a tropical country B. He has a rare disease from there. The hospital cannot recognize his disease. The clinic in country B records some cases like his. However, the two databases of the hospital and the clinic are not in the same database engine, or cloud provider. If a cloud federation exists to interconnect the two clouds, his disease could be recognized and he can have a treatment soon.

In cloud federations, pay-as-you-go models and elasticity thus raise an important issue in terms of Multi-Objective Optimization Problems (MOOPs) according to users preferences, such as time, money, quality, etc. However, MOOPs in a cloud federation are hard to solve due to issues of heterogeneity, and variability of the cloud environment, and high complexity in generating the space of candidate solutions.

Let's consider a query **Q** in a example below.

Example 1. A query **Q** in the medical domain, based on TPC-H[1] query 3 and 4:

```
SELECT p.UID, p.PatientID, s.PatientName,
p.PatientBrithDate, p.PatientSex,
p.EthnicGroup, p.SmokingStatus,
s.PatientAge, s.PatientWeight,
s.PatientSize, i.GeneralName,
i.GeneralValues, q.UID,
q.SequenceTags, q.SequenceVRs,
q.SequenceNames, q.SequenceValues
FROM Patient p, GeneralInfoTable i,
Study s, SequenceAttributes q
WHERE p.UID = s.UID AND p.UID = i.UID
AND p.UID = q.UID AND p.PatientSex = 'M'
AND p.SmokingStatus = 'NO' AND s.PatientAge >= x
AND q.SequenceNames
LIKE '%X-ray%'
```

---

[1] http://www.tpc.org/tpch/.

Table 1. Multiple Objectives for Query Execution Plans

QEP	VMs	Price ($/60 min)	Time (min)	Monetary ($)
QEP1	20	0.02	60	0.4
QEP2	80	0.02	22	0.59
QEP3	50	0.02	26	0.43

This query is transformed into a logical query plan using logical operation, such as *select, project, join, etc.* Depending on the physical operators, a query optimizer generates a query execution plan to execute a logical plan. Actually, various Query Execution Plans (QEPs) are generated with respect to the number of nodes, their capacity in terms of CPU, memory and disk and the pricing model. Table 1 presents an example of possible QEPs for **Q**. Choosing an execution plan is a trade-off between objectives such as the response time or the monetary cost, and depends on users' preferences: a user A may prefer minimizing his budget (QEP1); a user B may want the lowest response time (QEP2); a user C may look for a trade-off between time and money (QEP3).

Assuming that the query is processed on Amazon EC2. The master consists of a m2.4xlarge instance (8 virtual cores with 68.4 GB of RAM). Workers consist of m3.2xlarge instances (8 virtual cores and with 30 GB of RAM). If the pool of resources is 70 VCPU with 260 GB of memory, the number of QEPs is thus 70 × 260 = 18,200. The problem is then how to search and optimize such a query in a real environment, when the pool of resources is more variable, with respect to multiple dimensions (response time, monetary cost, etc.). Since generating QEPs maybe infeasible due to high complexity, we aim to find an approximate optimal solution.

In this paper, we address several challenges for the development of medical data management in cloud federations. The first challenge is how to estimate accurate parameter values for MOOPs without precise knowledge of the execution environment in a cloud federation consisting of different sites. The execution environment may consist of various hardware and systems. In addition, it also depends on the variety of physical machines, load evolution and wide-range communications. As a consequence, the estimation process is complex. The second challenge is how to find an approximate optimal solution in MOOPs using an efficient Multi-Objective Optimization algorithm. Indeed, MOOPs could be solved by Multi-Objective Optimization algorithms or the Weighted Sum Model (WSM) [24] or be converted to a Single-Objective Optimization Problem (SOOP). However, SOOPs cannot adequately represent MOOPs [23]. Also, MOOPs leads to find solutions by Pareto dominance techniques. Since generating a Pareto-optimal front is often infeasible due to high complexity [55], MOOPs need an approximate optimal solution calculated by Pareto dominance techniques.

The estimation process can be classified into two classes: without [39,42,50] and with machine learning algorithms [17]. In a cloud federation with variability and different systems, cost functions may be quite complex. In the first class, cost models introduced to build an optimal group of queries [39] are limited to MapReduce [12]. Besides, a PostgreSQL cost model [50] aims to predict query execution time for this specific relational DBMS. Moreover, OptEx [42] provides estimated job completion times for Spark[2] with respect to the size of the input dataset, the number of iterations, the number of nodes composing the underlying cloud. These works mention the estimation of only execution time for a job, and not for other metrics, such as monetary cost. Meanwhile, various machine learning techniques are applied to estimate execution time in recent research [2, 21, 46, 51]. They predict the execution time by many machine learning algorithms. They treat the database system as a black box and try to learn a query running time prediction model using the total information for training and testing in the model building process. It may lead to the use of expired information. In addition, most of these solutions solve the optimization problem with a scalar cost value and do not consider multi-objective problems.

A well known Pareto dominance technique to solve the high complexity of MOOP is Evolutionary Multiobjective Optimization (EMO). Among EMO approaches, Non-dominated Sorting Genetic Algorithms (NSGAs) [14, 15] have lower computational complexity than other EMO approaches [15]. However, these algorithms still have high computational complexity. We presented Non-dominated Sorting Genetic Algorithm based on Grid partitioning (NSGA-G) [36] to improve both computation time and qualities of NSGAs. It has more advantages than other NSGAs. Two versions of NSGA-G will be shown to compromise computation and quality.

In this paper, we introduce a medical system on a cloud federation called Medical Data Management System (MIDAS). It is based on the Intelligent Resource Scheduler (IReS) [17], an open source platform for complex analytics workflows executed over multi-engine environments. In particular, we focus on: (1) a dynamic estimation and (2) a Non-dominated Sorting Genetic Algorithm for Multi-Objective Optimization Problems. The first contribution is Dynamic linear REgression AlgorithM (DREAM) to provide accurate estimation with low computational cost. DREAM is then implemented and validated with experiments on a decision support benchmark (TPC-H benchmark). The second contribution is Non-dominated Sorting Genetic Algorithm based on Grid partitioning (NSGA-G) to improve both quality and computational efficiency of NSGAs, and also provides an alternative for Pareto-optimal of MOOPs. NSGA-Gs are validated through experiments on DTLZ problems [16] and compared with NSGA-II [15], NSGA-III [14], and the others in Generational Distance [49], Inverted Generational Distance [9], and Maximum Pareto Front Error [48] statistic.

This paper is an extended version of [36, 37]. In particular, they are grouped together to become a system, MIDAS. Besides, the theory of NSGA-G in [36] is expanded to two versions: NSGA-G using Min point and using Random metric. The remaining of this paper is organized as follows. Section 2 presents the research background. DREAM is presented in Sect. 3. Section 4 shows

---

[2] https://spark.apache.org/.

NSGA-G. Section 5 presents experiments to validate DREAM and NSGA-Gs. Finally, Sect. 6 concludes this paper and lists some perspectives.

2 Background

In this section, we introduce an architecture of the system, concepts and techniques, allowing us to implement the proposed medical data management on a cloud federation. First of all, an overview of the Medical Data Management System (**MIDAS**) and the benefits of cloud federation where our system is built on are introduced. After that, an open source platform, which helps our system managing and executing workflows over multi-engine environments is described. The concept of Pareto plan set related to Multi-Objective Optimization Problem (MOOP) in **MIDAS** is then defined. In addition, Multiple Linear Regression and Non-dominated Sorting Genetic Algorithm are also introduced as the basic foundation of our proposed algorithms for MOOP.

2.1 Cloud Federation

This section shows the definition and the example related to a cloud federation. A cloud federation enables to interconnect different cloud computing environments. Cloud computing [3] allows to access on demand and configurable resources, which can be quickly made available with minimal maintenance. According to the pay-as-you-go pricing model, customers only pay for resources (storage and computing) that they use. Cloud Service Providers (CSP) supply a pool of resources, development platforms and services. There are many CSPs on the market, such as Amazon, Google and Microsoft, etc., with different services and pricing models. For example, Table 2 shows the pricing of instances in two cloud providers in 2019. The price of Amazon instances are lower than the price of Microsoft instances, but the price of Amazon is without storage. Hence, depending on the demand of a query, the monetary cost is low or high at a specific provider.

In the medical domain, cloud federation may lead to query data across different clouds. A demand query running in that cloud federation could be concerned about the price of time and money of the execution query. It is a Multi-Objective Optimization Problem (MOOP). For example, federating resources makes it possible to access any information on a person with distributed hospital data on various sites. Various big data management system could be used to manage the medical data, which has the 3Vs characteristics of Big Data [1]: high volume, high variety, and high velocity. The data also stores that belong in different clouds are shown in Fig. 1. This example shows that the data can be stored in three different clouds, such as Amazon Web Services, Microsoft Azure, Google Cloud Platform. Pay-as-you-go models in clouds lead to solving Multi-Objective Optimization Problem to find a Query Execution Plan (QEP) according to users preferences, such as time, money, quality, etc. MOOPs often use Pareto dominance techniques in finding an optimal solution.

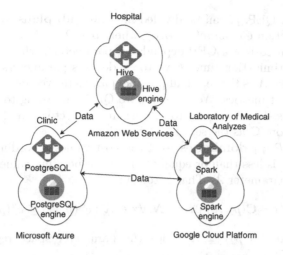

Fig. 1. Motivating Example on using cloud federation.

Table 2. Example of instances pricing in 2019.

Provider	Machine	vCPU	Memory (GiB)	Storage (GB)	Price ($/hour)
Amazon	a1.medium	1	2	EBS-Only	0.0255
	a1.large	2	4	EBS-Only	0.0510
	a1.xlarge	4	8	EBS-Only	0.1020
	a1.2xlarge	8	16	EBS-Only	0.2040
	a1.4xlarge	16	32	EBS-Only	0.4080
Microsoft	B1S	1	1	2	0.0104
	B1MS	1	2	4	0.0208
	B2S	2	4	8	0.0416
	B2MS	2	8	16	0.0832
	B4MS	4	16	32	0.1660
	B8MS	8	32	64	0.3330

2.2 Pareto Plan Set

Pareto dominance techniques are often used in Multi-Objective Optimization Problem (MOOP), such as Evolutionary Multiobjective Optimization (EMO) [14, 15,27,31,44,53,54]. In the vast space of candidate solutions of Multi-Objective Optimization Problem (MOOP), a candidate solution may be not better than another one because of trade-off between various objective values. Pareto sets are used in this situation to optimize a Multi-Objective Optimization Problem.

In particular, in a query processing problem, let a **query** q be an information request from databases, presented by a set of tables. A **Query Execution Plan** (QEP), denoted by p, is the evaluation of a query that can be passed to the executor. The set of QEPs is denoted by symbol $\mathcal{P}$. The set of operators is denoted by $\mathcal{O}$. A QEP, p, can be divided into two **sub-plans** p_1 and p_2 if p is the result of function $Combine(p_1, p_2, o)$, where $o \in \mathcal{O}$.

The execution cost of a QEP depends on parameters, which values are not known at the optimization time. A vector $\mathbf{x}$ denotes parameters value and the **parameter space** $\mathcal{X}$ is the set of all possible parameter vectors $\mathbf{x}$. N is denoted as the set of n cost metrics. We can compare QEPs according to n cost metrics which are processed with respect to the parameter vector $\mathbf{x}$ and cost functions $c^n(p, \mathbf{x})$. Let denote $\mathbf{C}$ as the set of cost function c.

Let $p_1, p_2 \in \mathcal{P}$, p_1 **dominates** p_2 if the cost values according to each cost metric of plan p_1 is less than or equal to the corresponding values of plan p_2 in all the space of parameter $\mathcal{X}$. That is to say:

$$\mathbf{C}(p_1, \mathcal{X}) \preceq \mathbf{C}(p_2, \mathcal{X}) \mid \forall n \in N, \forall \mathbf{x} \in \mathcal{X} : c^n(p_1, x) \leq c^n(p_2, x). \tag{1}$$

The function $Dom(p_1, p_2) \subseteq \mathcal{X}$ yields the parameter space region where p_1 dominates p_2 [47]:

$$Dom(p_1, p_2) = \{x \in \mathcal{X} \mid \forall n \in N : c^n(p_1, x) \leq c^n(p_2, x)\}. \tag{2}$$

Assume that in the area $\mathbf{x} \in \mathcal{A}, \mathcal{A} \subseteq \mathcal{X}$, p_1 dominates p_2, $\mathbf{C}(p_1, \mathcal{A}) \preceq \mathbf{C}(p_2, \mathcal{A})$, $Dom(p_1, p_2) = \mathcal{A} \subseteq \mathcal{X}$. p_1 **strictly dominates** p_2 if all values for the cost functions of p_1 are less than the corresponding values for p_2 [47], i.e.

$$StriDom(p_1, p_2) = \{x \in \mathcal{X} \mid \forall n \in N : c^n(p_1, x) < c^n(p_2, x)\}. \tag{3}$$

A **Pareto region** of a plan is a space of parameters where there is no alternative plan has lower cost than it [47]:

$$PaReg(p) = \mathcal{X} \setminus (\bigcup_{p^* \in \mathcal{P}} StriDom(p^*, p)). \tag{4}$$

2.3   IReS

Cloud federation model needs to integrate cloud services from multiple cloud providers. It raises an important issue in terms of heterogeneous database engines in various clouds. Among various heterogeneous database system described in Table 3, IReS platform considers both heterogeneous systems and Multi-Objective Optimization Problem in clouds.

Intelligent Multi-Engine Resource Scheduler (IReS) [17] is an open source platform for managing, executing and monitoring complex analytics workflows. IReS provides a method of optimizing cost-based workflows and customizable resource management of diverse execution and various storage engines. Especially, IReS platform helps us to organize data in the multiple clouds as a cloud federation. **Interface** is the first module which is designed to receive information on data and operators, as shown in Fig. 4. The second module is **Modelling**, as shown in Fig. 4, is used to predict the execution time by a model chosen by comparing machine learning algorithms. For example, Least squared

Table 3. Recent heterogeneous database system researches.

Research	Heterogeneous	MOOP
Proteus [28]	✓	✗
Polystore Query rewriting [40]	✓	✗
BigDAWG Polystore System [18]	✓	✗
ClooudMdsQL [32,33]	✓	✗
MuSQLE [22]	✓	✗
MISO [38]	✓	✗
Polybase [11]	✓	✗
Estoscada [6]	✓	✗
IReS	✓	✓

regression [41], Bagging predictors [5], Multilayer Perceptron in WEKA framework[3] are used to build the cost model in **Modelling** module. The module tests many algorithms and the best model with the smallest error is selected. It guarantees the predicted values as the best one for estimating process. Next module, **Multi-Objective Optimizer**, optimizes Multi-Objective Query Processing (MOQP) and generates a Pareto QEP set. In Multi-Objective problem, the objectives are the cost functions user concerned, such as the execution time, monetary, intermediate data, etc. Multi-Objective Optimization algorithms can be applied to the **Multi-Objective Optimizer**. For instance, the algorithms based on Pareto dominance techniques [10,14,15,27,31,36,44,53,54] are solutions for Multi-objective Optimization problems. Finally, the system selects the best QEP based on user query policy and Pareto set. The final QEP is run on multiple engines, as shown in Fig. 4.

2.4 Multiple Linear Regression

In many database management systems, predicting cost values is useful in optimization process [50]. Recent researches have been exploring the statistical machine learning approaches to build predictive models for this task. They often use historical data to train and test the cost model as a Single-Objective Problem (SOP). Besides, Linear Regression is an useful class of models in science and engineering [43]. In this section, we describe the background of this model.

This model is used in the situation in which a cost value, c, is a function of one or more independent variables $x_1, x_2, ..., $ and x_L. For example, execution time c is a function of data size x_1 of first element in join operator and data size x_2 of second element in that join operator.

Given a sample of c values with their associated values of x_i, $i = 1, 2, ..., L$. We focus in the estimation the relationship between c and the independent variables $x_1, x_2, ..., $ and x_L based on this sample. Cost function c of Multiple Linear Regression (MLR) model [43] is defined as follows:

---

[3] https://www.cs.waikato.ac.nz/ml/weka/.

$$c = \beta_0 + \beta_1 x_1 + \dots + \beta_L x_L + \epsilon, \tag{5}$$

where β_l, $l = 0, \dots, L$, are unknown coefficients, $x_l, l = 1, \dots, L$, are the independent variables, e.g., size of data, computer configuration, etc., c is cost function values and ϵ is random error following normal distribution $\mathcal{N}(0, \sigma^2)$ with zero mean and variance σ^2. The **fitted equation** is defined by:

$$\hat{c} = \hat{\beta}_0 + \hat{\beta}_1 x_1 + \dots + \hat{\beta}_L x_L. \tag{6}$$

Example 2. A query **Q** [37] could be expressed as follows:

```
SELECT p.PatientSex, i.GeneralNames
FROM Patient p, GeneralInfo i
WHERE p.UID = i.UID
```

where Patient table is stored in cloud A and uses Hive [45] database engine[4], while GeneralInfo table is in cloud B with PostgreSQL database engine[5]. This scenario leads to concern two metrics of monetary cost and execution time cost. We can use the cost functions which depend on the size of tables of Patient and GeneralInfo. Besides, the configuration and pricing of virtual machines cloud A and B are different. Hence, the cost functions depend on the size of tables and the number of virtual machines in cloud A and B.

$$\hat{c}^{ti} = \hat{\beta}_{t0} + \hat{\beta}_{t1} x_{Pa} + \hat{\beta}_{t2} x_{Ge} + \hat{\beta}_{t3} x_{nodeA} + \hat{\beta}_{t4} x_{nodeB}$$

$$\hat{c}^{mo} = \hat{\beta}_{m0} + \hat{\beta}_{m1} x_{Pa} + \hat{\beta}_{m2} x_{Ge} + \hat{\beta}_{m3} x_{nodeA} + \hat{\beta}_{m4} x_{nodeB}$$

where $\hat{c}^{ti}, \hat{c}^{mo}$ are execution time and monetary cost function; x_{Pa}, x_{Ge} are the size of Patient and GeneralInfo tables, respectively, and x_{nodeA}, x_{nodeB} are the number of virtual machines created to run query **Q**.

There are M historical data, each of them associates with a response c_m, which can be predicted by a **fitted value** $\hat{c}_m$ calculated from corresponding x_{lm} as follows:

$$\hat{c}_m = \hat{\beta}_0 + \hat{\beta}_1 x_{1m} + \dots + \hat{\beta}_L x_{Lm}; m = 1, \dots, M. \tag{7}$$

Let denote

$$A = \begin{bmatrix} 1 & x_{11} & x_{21} & \dots & x_{L1} \\ 1 & x_{12} & x_{22} & \dots & x_{L2} \\ . & . & & . & . \\ . & . & & . & . \\ 1 & x_{1M} & x_{2M} & \dots & x_{LM} \end{bmatrix}, \tag{8}$$

$$C = \begin{bmatrix} c_1 \\ c_2 \\ . \\ . \\ c_M \end{bmatrix}, \tag{9}$$

---

[4] http://hive.apache.org/.
[5] https://www.postgresql.org/.

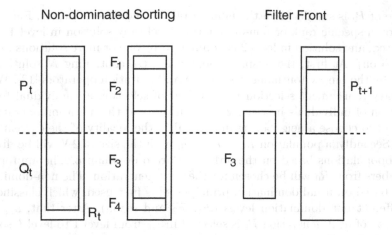

Fig. 2. NSGA-II and NSGA-III procedure [14,15].

$$B = \begin{bmatrix} \hat{\beta}_0 \\ \hat{\beta}_1 \\ \vdots \\ \hat{\beta}_L \end{bmatrix}. \tag{10}$$

To minimize the **Sum Square Error** (SSE), defined by:

$$SSE = \sum_{m=1}^{M} (c_m - \hat{c}_m)^2, \tag{11}$$

the solution for B is retrieved by:

$$B = (A^T A)^{-1} A^T C. \tag{12}$$

2.5 NSGA

After having the prediction cost values of MOOPs, we need to use Multi-Objective Optimization algorithms to find an optimal solution.

Among Multi-objective Optimization algorithm classes, Evolutionary Multi-objective Optimization (EMO) shows their advantages in searching and optimizing for the MOOPs [10]. Among EMO approaches, Non-dominated Sorting Genetic Algorithms provide low computational complexity of non-dominated sorting, $O(MN^2)$ of NSGAs [14,15] comparing to $O(MN^3)$ of other Evolutionary Multi-Objective Optimization (EMO), where M is the number of objectives and N is the population size.

NSGA Process. Initially, NSGAs start with a population P_0 consisting of N solutions. In MOOPs, a population represents a set of candidate solutions.

The size of P_0 is smaller than the number of all candidate solutions. Each solution is on a specific rank or non-domination level (any solution in level 1 is not dominated, any solution in level 2 is dominated by one or more solutions in level 1 and so on). At first, the offspring population Q_0 containing N solutions, is created by the binary tournament selection and mutation operators [13]. Where the binary tournament selection is a method of selecting an individual from a population of individuals in a genetic algorithm, and the mutation operation is a method to choose a neighboring individual in the locality of the current individual. Secondly, a population $R_0 = P_0 \cup Q_0$ with the size of $2N$ will be divided into subpopulations based on the order of Pareto dominance. The appropriate N members from R_0 will be chosen for the next generation. The non-dominated sorting based on usual domination principle [8] is first used, which classifies R_0 into different non-domination levels ($\mathcal{F}_1$, $\mathcal{F}_2$ and so on). After that, a parent population of next-generation P_1 is selected in R_0 from level 1 to level k so that the size of $P_1 = N$ and so on.

The difference among NSGA-II, NSGA-III and other NSGAs is the way to select members in the last level $\mathcal{F}_l$. To keep the diversity, NSGA-II [15] and SPEA-II [54] use crowding distance among solutions in their selection. NSGA-II procedure is not suitable for MOO problems and the crowding distance operator needs to be replaced for better performance [26, 34]. Hence, when the population has a high-density area, higher than others, NSGA-II prefers the solution which is located in a less crowded region.

On the other hand, MOEA/D [53] decomposes a multiple objectives problem into various scalar optimization subproblems. The diversity of solutions depends on the scalar objectives. However, the number of neighborhoods needs to be declared before running the algorithm. In addition, the estimation of good neighborhood is not mentioned. The diversity is considered as the selected solution associated with these different sub-problems. Experimental results in [14] show various versions of MOEA/D approaches which fail to maintain a good distribution of points.

An Evolutionary Many-Objective Optimization Algorithm Using Reference-point Based Non-Dominated Sorting Approach [14] (NSGA-III) uses different directions to maintain the diversity of solutions. NSGA-III replaces the crowding distance operator by comparing solutions. Each solution is associated to a reference point [14], which impacts the execution time to built the reference points in each generation. The diversity of NSGA-III is better than the others, but the execution time is very high. For instance, with two objectives and two divisions, three reference points will be created, (0.0, 1.0), (1.0, 0.0) and (0.5, 0.5), as shown in Fig. 3. After selection process, the diversity of population is better than NSGA-II with solutions close to three reference points. However, comparing all solutions to each reference point makes the computation time of NSGA-III very high.

In addition, NSGAs often compare all solutions to choose good solutions in $\mathcal{F}_l$. Therefore, when the number of solutions or objectives is significant, the time for calculating and comparing is considerable.

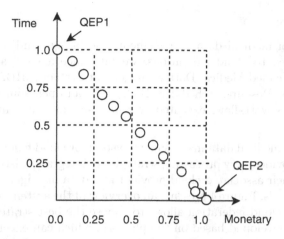

Fig. 3. An example of using the crowing distance in NSGA-II.

Application. In some cases, some objectives are homogeneous. In the reason of the homogeneity between the multi-objectives functions, removing an objective do not affect to the final results of MOO problem. In other cases, the objectives may be contradictory. For example, the monetary is proportional to the execution time in the same virtual machine configuration in a cloud. However, cloud providers usually leases computing resources that are typically charged based on a per time quantum pricing scheme [30]. The solutions represent the trade-offs between time and money. Hence, the execution time and the monetary cost cannot be homogeneous.

As a consequence, the multi-objective problem cannot be reduced to a mono-objective problem. Moreover, if we want to reduce the MOO to a mono-objective optimization, we should have a policy to group all objectives by the Weighted Sum Model (WSM) [24]. However, estimating the weights corresponding to different objectives in this model is also a multi-objective problem.

In addition, MOO problems could be solved by MOO algorithms or WSM [24]. However, MOO algorithms are selected thanks to their advantages when comparing with WSM. The optimal solution of WSM could be unacceptable, because of an inappropriate setting of the coefficients [20]. Furthermore, the research in [29] proves that a small change in weights may result in significant changes in the objective vectors and significantly different weights may produce nearly similar objective vectors. Moreover, if WSM changes, a new optimization process will be required.

In conclusion, MOOP approaches leads to using Pareto dominance techniques. A pareto-optimal front is often infeasible [55]. NSGAs show the advantage in searching a Pareto solution for MOOP in less computational complexity than other EMO [15]. However, they should be improved the quality to solve MOOP when the number of objectives is significant.

2.6　Motivation

In the context of medical data management, the background of concepts and techniques related to cloud federations, we introduce a medical system on a cloud federation called Medical Data Management System (MIDAS). It is based on the Intelligent Resource Scheduler (IReS) [17], an open source platform for complex analytics work-flows executed over multi-engine environments.

MIDAS. It is a medical data management system for cloud federation. The proposal aims to provide query processing strategies to integrate existing information systems (with their associated cloud provider and data management system) for clinics and hospitals. Figure 4 presents an overview of the system. Integrating the system within a cloud federation allows to choose the best strategy for MOQP. MIDAS can be developed based on the platform which can execute over multi-engine environments on clouds. Figure 4 also shows an example of MIDAS, where three database engines are installed and run in three clouds of different providers.

We choose IReS platform to consider the advantage as shown in Table 3. IReS platform is installed in every machine in MIDAS. The different cloud resource pools allow the system to run in the most appropriate infrastructure environments. The system can optimize workflows between different data sources on different clouds, such as Amazon Web Services[6], Microsoft Azure[7] and Google Cloud Platform[8]. The proposed system is developed based on the Intelligent Resource Scheduler (IReS) for complex analytics workflows executed over multi-engine environments on a cloud federation.

Machine Learning Algorithm. The machine learning algorithms in IReS need entire training datasets to estimate the running costs, which are calculated by determining the cost of processing a job. It may lead to use expired information. Hence, the proposal algorithm aims to improve the accuracy of estimated values with low computational cost. Our proposed method is integrated into IReS to predict the cost values with low computational cost in MOQP of a cloud environment.

Multi-Objective Optimization. In addition, MOQP could be solved by Multi-Objective Optimization algorithms or the Weighted Sum Model (WSM) [24]. However, Multi-Objective Optimization algorithms may be selected thanks to their advantages when comparing with WSM. The optimal solution of WSM could be not acceptable, because of an inappropriate setting of the coefficients [20]. Furthermore, the research in [29] proves that a small change in weights may result in significant changes in the objective vectors and significantly different weights may produce nearly similar objective vectors. Moreover, if WSM changes, a new optimization process will be required. Hence, our system applies a Multi-Objective Optimization algorithm to the **Multi-Objective**

---

[6] https://aws.amazon.com/.
[7] https://azure.microsoft.com/.
[8] https://cloud.google.com/.

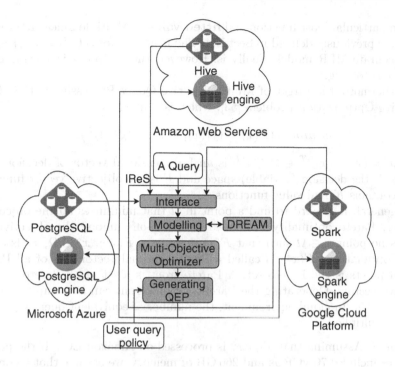

Fig. 4. Architecture of MIDAS [37].

Optimizer to find a Pareto-optimal solution. When the WSM changes, the final result just is determined by using the Pareto-optimal set at the final step.

Furthermore, generating a Pareto-optimal front is often infeasible due to high complexity [55]. MOOPs leads to finding an approximate optimal solution by Pareto dominance techniques. A well known approach to solve the high complexity of MOOP is Evolutionary Multiobjective Optimization (EMO). Among EMO approaches, Non-dominated Sorting Genetic Algorithms (NSGAs) [14,15] have lower computational complexity than other EMO approaches [15]. However, this algorithm still has high computational complexity. We need to find an approach to improve the computational complexity and quality of NSGAs.

In conclusion, our solutions aim to improve the accuracy of cost value prediction with low computational cost and to solve MOQP by Multi-Objective Optimization algorithm in a cloud federation environment. Besides, we also find a method to search and optimize MOOPs by finding an approximate optimal solution in the high complexity of generating a Pareto-optimal front.

3   Dynamic Regression Algorithm

The first technique in MIDAS relates to the estimation of accurate cost values in the variable environment of a cloud federation. Most of cost models [19,39,50] depend on the size of data. Hence, our cost functions are functions of the size of data. In particular, cost function and **fitted value** of Multiple Linear Regression model are previously defined in Sect. 2.4. The bigger M for sets $\{c_m, x_{lm}\}$ is, the more accurate MLR model usually is. However, the computer is slowing down when M is too big.

Furthermore, the target of Multi-Objective Query Processing is the Multi-Objective Optimization Problem [53], which is defined by:

$$minimize(F(x) = (f_1(x), f_2(x)..., f_K(x))^T), \tag{13}$$

where $x = (x_1, ..., x_L)^T \in \Omega \subseteq \mathbb{R}^L$ is an L-dimensional vector of decision variables, Ω is the decision (variable) space and F is the objective vector function, which contains K real value functions.

In general, it is hard to find a point in Ω that minimizes all the objectives together. Pareto optimality is defined by trade-offs among the objectives. If there is no point $x \in \Omega$ such that $F(x)$ dominates $F(x^*)$, $x^* \in \Omega$, x^* is called Pareto optimal and $F(x^*)$ is called a Pareto optimal vector. Set of all Pareto optimal points is the Pareto set. A Pareto front is a set of all Pareto optimal objective vectors. Generating the Pareto-optimal front can be computationally expensive [55]. In cloud environment, the number of equivalent query execution plans is multiplied.

Example 3. Assuming that a query is processed on Amazon EC2. If the pool of resources includes 70 vCPUs and 260 GB of memory, we assume that a configuration to execute a query plan is created by the number of vCPUs and the size of memory which is the multiple of 1 GB. In particular, a configuration can be 01 vCPU and 260 GB of memory and the others is 70 vCPUs and 01 GB of memory. Hence, the combination of different configurations to execute this query would be 70 * 260 = 18,200.

Example 3 shows that a query plan can generate multiple equivalent QEPs in cloud environment. The smaller M for sets $\{c_m, x_{lm}\}$ is, the faster speed for the estimation cost process of Multi-Objective Query Processing for a QEP is. In the system of computationally expensiveness in cloud environment as in Example 3, a small reduction of computation for an equivalent QEP estimation will become significant for a large number of equivalent QEPs estimation.

The most important idea is to estimate MLR quality by using the coefficient of determination. The coefficient of determination [43] is defined by:

$$R^2 = 1 - SSE/SST, \tag{14}$$

where SSE is the sum of squared errors and SST represents the amount of total variation corresponding to the predictor variable X. Hence, R^2 shows the proportion of variation in cost given by the Multiple Linear Regression model of variable X. For example, the model gives $R^2 = 0.75$ of time response cost, it can be concluded that 3/4 of the variation in time response values can be explained by the linear relationship between the input variables and time response cost. Table 4 presents an example of MLR with different number of measures. The smallest dataset is $M = L + 2 = 4$ [43], where M is the size of previous

Table 4. Using MLR in different size of dataset [37].

Cost	x_1	x_2	M	R^2
20.640	0.4916	0.2977		
15.557	0.6313	0.0482		
20.971	0.9481	0.8232		
24.878	0.4855	2.7056	4	0.7571
23.274	0.0125	2.7268	5	0.7705
30.216	0.9029	2.6456	6	0.8371
29.978	0.7233	3.0640	7	0.8788
31.702	0.8749	4.2847	8	0.8876
20.860	0.3354	2.1082	9	0.8751
32.836	0.8521	4.8217	10	0.8945

data and L is the number of variables in Eq. (5). In general, R^2 increases in parallel with M. In particular, R^2 should be greater than 0.8 to provide a sufficient quality of service level. As a consequence, M should be greater than 5 to provide enough accuracy. Hence, when the system requires the minimum values of R^2 is equal to 0.8, $M > 6$ is not recommended. In general, R^2 still rises up when M goes up. Therefore, we need to determine the model which is sufficient suitable by the coefficient of determination.

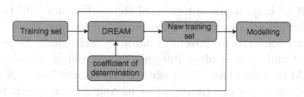

Fig. 5. DREAM module [37].

Our motivation is to provide accurate estimation while reducing the number of previous measures based on R^2. We thus propose DREAM as a solution for cloud federation and their inherent variance, as shown in Fig. 5. DREAM uses the training set to test the size of new training dataset. It depends on the predefined coefficient of determination. The new training set is generated in order to have the updated value and avoid using the expired information. With the new training set, **Modelling** uses fewer data in the building model process than the original approach does.

Cost modeling without machine learning [39, 42, 50] often uses the size of data to estimate the execution time for the specific system. Besides, the machine learning approach [17] can use any information to estimate the cost value. Hence, our algorithm uses the size of data as variables of DREAM. In (6), $\hat{c}$ is the cost value, which needs to be estimated in MOQP, and $x_1, x_2, \ldots$ are the information of system, such as size of input data, the number of nodes, the type of virtual

---

Algorithm 1. Calculate the predict value of multi-cost function [37]

1: **function** ESTIMATECOSTVALUE($\mathbf{R}^2_{require}, X, M_{max}$)
2:     **for** $n = 1$ to N **do**
3:         $\mathbf{R}^2_n \leftarrow \emptyset$ //with all cost function
4:     **end for**
5:     $m = L + 2$ //at least $m = L + 2$
6:     **while** (any $R^2_n < R^2_{n-require}$) and $m < M_{max}$ **do**
7:         **for** $\hat{c}_n(p) \subseteq \hat{\mathbf{c}}_\mathbf{N}(\mathbf{p})$ **do**
8:             $R^2_n = 1 - SSE/SST$
9:             $\hat{c}_n = \hat{\beta}_{n0} + \hat{\beta}_{n1}x_1 + ... + \hat{\beta}_{nL}x_L$
10:        **end for**
11:        $m = m + 1$
12:    **end while**
13:    **return** $\hat{\mathbf{c}}_\mathbf{N}(\mathbf{p})$
14: **end function**

---

machines. If $R^2 \geq R^2_{require}$, where $R^2_{requires}$ is predefined by users, the model is reliable. In contrast, it is necessary to increase the number of set value. Algorithm 1 shows a scheme as an example of increasing value set: $m = m + 1$.

In this paper, we focus on the accuracy of execution time estimation with the low computational cost in MOQP. The original optimization approach in IReS uses Weighted Sum Model (WSM) [24] with user policy to find the best candidate solution. However, the optimal solution of WSM could be not acceptable, because of an inappropriate setting of the coefficients [20]. Besides, Multi-Objective Optimization algorithms have more advantages than WSM [20,29]. They lead to find solutions by Pareto dominance techniques. However, generating a Pareto-optimal front is often infeasible due to high complexity [55]. One of well known Multi-Objective Optimization algorithm class is Non-dominated Sorting Algorithms (NSGAs). Hence, after having a set of predicted cost function values for each query plan, a Multi-Objective Optimization algorithm, such as NSGA-G [36] is applied to determine a Pareto query execution plan set. At the final step, the weight sum model $\mathbf{S}$ and the constraint $\mathbf{B}$ associated with the user policy are used to return the best QEP for the given query [24]. In particular, the most meaningful plan will be selected by comparing function values with weight parameters between $\hat{c}_n$ [24] at the final step, as shown in Algorithm 2. Figure 6 shows the different between two MOQP approaches.

4   Non-dominated Sorting Genetic Algorithm Based on Grid Partitioning

After having the prediction cost values of MOOPs by DREAM, we need to use Multi-Objective Optimization algorithms to find an optimal solution. Hence, the second technique relates to looking for an efficient approach for searching and optimizing in MIDAS is introduced in this section. NSGAs [14,15] are well

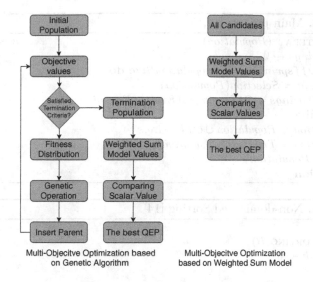

Fig. 6. Comparing two MOQP approaches [37]

Algorithm 2. Select the best query plan in $\mathcal{P}$ [37]

1: **function** BESTINPARETO($\mathcal{P}, \mathbf{S}, \mathbf{B}$)
2:     $P_B \leftarrow p \in \mathcal{P} | \forall n \leq |\mathbf{B}| : c_n(p) \leq B_n$
3:     **if** $P_B \neq \emptyset$ **then**
4:         **return** $p \in P_B | C(p) = min(WeightSum(P_B, \mathbf{S}))$
5:     **else**
6:         **return** $p \in \mathcal{P} | C(p) = min(WeightSum(\mathcal{P}, \mathbf{S}))$
7:     **end if**
8: **end function**

---

known approaches to optimize MOOPs. Our previous work [36] proposed Non-dominated Sorting Genetic Algorithm based on Grid partitioning (NSGA-G) to improve both diversity and convergence of NSGAs while having an efficient computation time by reducing the space of selected good solutions in the truncating process. NSGA-G is an algorithm based on genetic algorithms (GAs). The convergence of GAs is discussed in [7]. The difference between many GAs is the qualities of diversity and convergence. We will describe the strategy to improve the qualities of NSGAs while having an efficient computation time as below.

At the tth generation of Non-dominated Sorting Genetic Algorithms, P_t represents the parent population with N size and Q_t is offspring population with N members created by P_t. $R_t = P_t \cup Q_t$ is a group in which N members will be selected for P_{t+1}.

Our algorithms are developed based on the MLR described above using x_i for size of data and c_i for the metric cost, such as the execution time, energy consumption, etc.

Algorithm 3. Main process [14,15].

1: **function** ITERATE($Population$)
2:     $Offsprings \leftarrow \emptyset$
3:     **while** $Offsprings.size < populationSize$ **do**
4:         $Parent = Selection(Population)$
5:         $Offsprings = Offsprings \cup Evolve(Parent)$
6:     **end while**
7:     $Population = Population \cup Offsprings$
8:     $Population = Truncate(Population)$
9:     **return** $Population$
10: **end function**

Algorithm 4. Non-dominated Sorting [14].

Require: R
1: **function** SORTING(R)
2:     $RinRank \leftarrow \emptyset$
3:     $rank = 1$
4:     $remaining \leftarrow R$
5:     **while** $RisNotEmpty$ **do**
6:         $Front \leftarrow non - dominatedPopulation(remaining, rank)$
7:         $remaining = remaining \setminus Front$
8:         $RinRank = RinRank \cup Front$
9:         $rank + +$
10:     **end while**
11:     **return** $RinRank$
12: **end function**

4.1 Main Process

This section describes more details about the main process of NSGAs. Algorithm 3 shows the steps of the processing. First, the Offspring is initialized in Line 2. The size of Offspring equals to the size of Population, i.e., N. Hence, a parent is selected from the population and evolved to become a new offspring. A new population with the size of $2N$ is created from Offspring and the old population. After that, the function Truncate will cut off the new population to reduce the members to the size of N, as shown in Line 8.

4.2 Non-dominated Sorting

Before the truncating process, the solutions in the population with a size of $2N$ should be sorted in multiple fronts with their ranking, as shown in Algorithm 4. First, the Non-dominated sorting operator generates the first Pareto set in a population of $2N$ solutions. Its rank is 1. After that, the process is repeated until the remain population is empty. Finally, $2N$ solutions are divided into various fronts with their ranks.

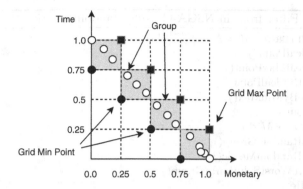

Fig. 7. An example of using Grid points.

Algorithm 5. Filter front in NSGA-G using Min point. [36]

1: **function** FILTER($\mathcal{F}_l, M = N - \sum_{j=1}^{l-1} \mathcal{F}_j$)
2:     updateIdealPoint()
3:     updateIdealMaxPoint()
4:     translateByIdealPoint()
5:     normalizeByMinMax()
6:     createGroups
7:     **while** $| \mathcal{F}_l | > M$ **do**
8:         selectRandomGroup()
9:         removeMaxSolutionInGroup()
10:     **end while**
11:     **return** $\mathcal{F}_l$
12: **end function**

4.3   Filter Front Process

NSGA-G Using Min Point. NSGA-G finds the nearest smaller and bigger grid point for each solution. For example, Fig. 7 shows an example of a two-objective problem. If the unit of the grid point is 0.25 (the size of grid is 4) and the solution with two-objective value is $[0.35, 0.65]$, the closest Grid Min Point is $[0.25, 0.5]$ and the nearest Grid Max Point is $[0.5, 0.75]$.

The first strategy avoids computing multiple objective cost values of all solutions in the population, the space is divided into multiple small groups by Grid Min Point and Grid Max Point, as shown in Fig. 7. Each group has one Grid Min Point, the nearest smaller point and one Grid Max Point, the nearest bigger point. Only solutions in a group are calculated and compared. The solution has the smallest distance to the nearest smaller point in a group will be added to P_{t+1}.

In this way, in any loop, we do not need to calculate the crowding-distance values or estimate the smallest distance from solutions to the reference points among all members in the last front, as shown in Fig. 7. In any loop, it is not necessary to compare solutions among all members in F_l, as F_3 in Fig. 2. The second strategy chooses randomly a group. The characteristic of diversity is maintained by this

---

Algorithm 6. Filter front in NSGA-G using Random metric.

1: **function** FILTER($\mathcal{F}_l, M = N - \sum_{j=1}^{l-1} \mathcal{F}_j$)
2:     updateIdealPoint()
3:     updateIdealMaxPoint()
4:     translateByIdealPoint()
5:     normalizeByMinMax()
6:     createGroups
7:     **while** $| \mathcal{F}_l | > M$ **do**
8:         selectRandomGroup()
9:         selectRandomMetric()
10:         removeWorstSolutionInGroup()
11:     **end while**
12:     **return** $\mathcal{F}_l$
13: **end function**

---

strategy. Both strategies are proposed to improve the qualities of our algorithm. Algorithm 5 shows the strategy to select $N - \sum_{j=1}^{l-1} \mathcal{F}_j$ members in $\mathcal{F}_l$.

The two lines 2 and 3 in Algorithm 5 determine the new origin coordinates and the maximum objective values of all solutions, respectively. After that, they will be normalized in a range of $[0, 1]$. All solutions will be in different groups, depending on the coefficient of the grid. The most important characteristic of this algorithm is randomly selecting the group like NSGA-III to keep the diversity characteristic and remove the solution among members of that group. This selection helps to avoid comparing and calculating the maximum objectives in all solutions.

To estimate the quality of the proposed algorithm, three qualities, Generational Distance [49], Inverted Generational Distance [9] and the Maximum Pareto Front Error [48], are used.

NSGA-G Using Random Metric. In MOOP, when the number of objectives is significant, any function used to compare solutions leads to high computation. NSGA-G using Min point uses Grid partition to reduce the number in groups, but it still needs a function to group all objectives value to a scalar value. In order to decrease the execution time, this section proposes a random method to compare solutions among a group. This approach does not generate any reference point or an intermediate function to estimate the value of solutions. The natural metric values are chosen randomly to remove the worst solution in the different groups.

All the steps in this algorithm are similar to NSGA-G using Min point, as shown from Line 2 to Line 6. Loop *While* has one more step of choosing metric randomly. Function *selectRandomMetric* is used to select a natural metric among the objectives in MOOP. The important characteristics of this algorithm are randomly selecting the group like NSGA-G and using natural metric among various objectives. It aims to keep the diversity characteristic, and reduce the comparing time. This selection helps to avoid using an intermediate function in comparing and calculating the values of solutions.

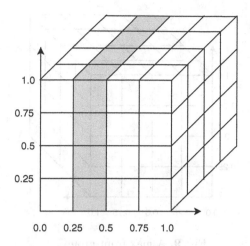

Fig. 8. A simple front group.

4.4 Selecting the Size of Grid

The proposed approach uses Grid partitioning to guarantee that the solutions are distributed in all the solution space. Assuming that there is a problem with N objectives. The last front should remove k solutions. By normalizing the space of solution in the range of $[0, 1]$ and dividing that range to n segments, a solution belongs to one of n^N groups in that space. In terms of Non-dominated principle, a group including a solution in that space have many other groups which contain Non-dominated solutions. These groups are called *Non-dominated groups*. All the groups in this situation make a set groups, called *front group*.

The proposed idea is to keep the diversity characteristic of the genetic algorithm by generating k groups and removing k solutions. Hence, the ideal *front group* is designed so that it has k groups.

Simple Front Group. From a group in the normalizing space in range of $[0, 1]$, a simple plane covers it and includes *Non-dominated groups*. In the space of N axes, the number of groups is n^N. Hence, the simple *front group* is the simple plane. The number of groups in that *front group* is n^{N-1}. Therefore, if the last front needs to remove k solutions, the number of grid n is determined as follows

$$n = \lceil k^{\frac{1}{N-1}} \rceil. \tag{15}$$

For example, Fig. 8 shows a problem with 3 objectives. In each axis coordinate, the size of grid is 4, and the maximum number of groups in all space of N axis coordinates is 4^3. A simple *front group* includes $4^{3-1} = 16$ groups. If the last front needs to remove 15 solutions, the number of grid when we choose simple front group is $n = \lceil k^{\frac{1}{N-1}} \rceil = 4$.

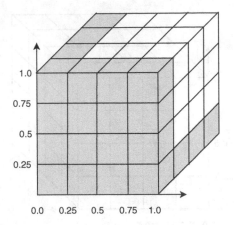

Fig. 9. A max front group.

Max Front Group. From a group in the normalizing space in range of $[0, 1]$, a simple plane covers it and includes *Non-dominated groups*. In the space of N axis coordinates, the number of groups is n^N. *Max front group* has the largest number of groups includes N planes. Hence, the number of groups in *Max front group* is $n^N - (n - 1)^N$. Therefore, if the last front needs to remove k solution, the number of grid n is determined as follows

$$n^N - (n - 1)^N = k. \tag{16}$$

For instance, Fig. 9 shows a problems with 3 objectives. In each axis coordinate, the size of grid is 4, the maximum number of groups in all space of N axis coordinates is 4^3. *Max front group* includes $4^3 - 3^3 = 64 - 27 = 37$ groups.

5   Validation

5.1   DREAM

The previous section introduces two algorithms for the Multi-Objective Optimization Problem in MIDAS. DREAM and NSGA-G have been implemented on top of IReS platform. They have been validated with experiments.

Implementation. Our experiments are executed on Galactica private cloud[9] with a cluster of three machines. Each node has four 2.4 GHz CPU, 80 GiB Disk, 8 GiB memory and runs 64-bit platform Linux Ubuntu 16.04.2 LTS. The system uses Hadoop 2.7.3[10], Hive 2.1.1, PostgreSQL 9.5.14, Spark 2.2.0 and Java OpenJDK Runtime Environment 1.8.0. IReS platform is used to manage data in multiple database engine and deploy the algorithms.

---

[9] https://horizon.isima.fr/.
[10] http://hadoop.apache.org/.

Table 5. Comparison of mean relative error with 100MiB TPC-H dataset [37].

Query	BML$_N$	BML$_{2N}$	BML$_{3N}$	BML	DREAM
12	0.265	0.459	0.220	0.485	0.146
13	0.434	0.517	0.381	0.358	0.258
14	0.373	0.340	0.335	0.358	0.319
17	0.404	0.396	0.267	0.965	0.119

Table 6. Comparison of mean relative error with 1GiB TPC-H dataset [37].

Query	BML$_N$	BML$_{2N}$	BML$_{3N}$	BML	DREAM
12	0.349	0.854	0.341	0.480	0.335
13	0.396	0.843	0.457	0.487	0.349
14	0.468	0.664	0.539	0.790	0.318
17	0.620	0.611	0.681	0.970	0.536

Experiments. TPC-H benchmark with two datasets of 100 MB and 1 GB is used to have experiments with DREAM. Experiments with TPC-H benchmark are executed in a multi-engine environment consisting of Hive and PostgreSQL deployed on Galactica private cloud. In TPC-H benchmark, the queries related to two tables are 12, 13, 14 and 17. These queries with two tables in two different databases, such as Hive and PostgreSQL, are studied.

Results. To estimate the quality of price models which are estimated by DREAM in comparison with other algorithms, Mean Relative Error (MRE), a metric used in [2] is used and described as below:

$$\frac{1}{M} \sum_{M}^{i=1} \frac{|\hat{c}_i - c_i|}{c_i}, \tag{17}$$

where M is the number of testing queries, $\hat{c}_i$ and c_i are the predict and actual execution time of testing queries, respectively. IReS platform uses multiple machine learning algorithms in their model, such as Least squared regression, Bagging predictors, Multilayer Perceptron.

In IReS model building process, IReS tests many algorithms and the best model with the smallest error is selected. It guarantees the predicted values as the best one for estimating process. DREAM is compared to the Best Machine Learning model (BML) in IReS platform with many observation window (N, $2N$, $3N$ and no limit of history data). The smallest size of a window, $N = L + 2$ [43], where L is the number of variables, is the minimum data set DREAM requires.

As shown in Table 5 and 6, MRE of DREAM are the smallest values between various observation windows. In our experiments, the size of historical data, which DREAM uses, are small, around N.

5.2 NSGA-G

Various earlier studies on Multiple Objective Evolutionary Algorithms (MOEAs) introduce test problems which are either simple or not scalable. DTLZ test problems [16] are useful in various research activities on MOEAs, e.g., testing the performance of a new MOEA, comparing different MOEAs and a better understanding of MOEAs. The proposed algorithm is experimented on DTLZ test problems with other famous NSGAs to show advantages in convergence, diversity and execution time.

Implementation. Our experiments use Multiobjective Evolutionary Algorithms (MOEA)[11] framework in Open JDK Java 1.8. All experiments are run on a machine with following parameters: Intel(R) core(TM) i7-6600U CPU @ 2.60 GHz × 4, 16 GB RAM.

Experiments. For fair comparison and evaluation, the same parameters are used, such as Simulated binary crossover [13] (30), Polynomial mutation [13] (20), max evaluations (10000) and populations (100) for eMOEA [10], NSGA-II, MOEA/D [53], NSGA-III and NSGA-G[12], during 50 independent running to solve two types of problems in DTLZ test problems [16] with m objectives, $m \in [5, 10]$. These algorithms use the same population size $N = 100$ and the maximum evaluation $M = 10000$. We apply *Simple front group* approach, Eq. 15, to determine the grid in both of NSGA-Gs with Min point and Random metric experiments. We use the Generational Distance (GD) [49], Inverted Generational Distance (IGD) [9] and the Maximum Pareto Front Error (MPFE) [48] to compare the quality of NSGA-Gs to other NSGAs.

GD measures how far the evolved solution set is from the true Pareto front [52], as shown in following:

$$GD = \frac{\sqrt{\sum_{i=1}^{n} d_i^2}}{n},$$ (18)

where $d_j = \min_j ||f(x_i) - PF_{true}(x_j)||$ shows the distance objective space between solution x_i and the nearest member in the true Pareto front (PF_{true}), and n is the number of solutions in the approximation front. Lower value of GD represents a better quality of an algorithm.

IGD is a metric to estimate the approximation quality of the Pareto front obtained by MOO algorithms [4], which can measure both convergence and diversity in a sense. IGD is shown in the following equation [52]:

$$IGD = \frac{\sum_{v \in PF_{true}} d(v, X)}{|PF_{true}|},$$ (19)

---

[11] http://moeaframework.org/.
[12] https://gitlab.inria.fr/trle/moea.

where X is the set of non-dominated solutions in the approximation front, $d(v, X)$ presents the minimum Euclidean distance between a point v in PF_{true} and the points in X. Lower value of IDG represents the approximate front getting close to PF_{true}, and not missing any part of the whole PF_{true}.

MPFE shows the most significant distance between the individuals in Pareto front and the solutions in the approximation front [52]. This metric is shown in the following equation:

$$MPFE = \max_i d_i. \tag{20}$$

In all tables show the experiments, the darkest mark value show the least value in various algorithm experiments, and the brighter mark value is the second least value among them.

Study on Test Problems. In this section, we use DTLZs, and WFG [25] test problem to experiment NSGA-Gs. Advantages of two versions of NSGA-G are present in Table 7, 8, 9, 10, 11, and 12. Metrics, such as GD, IDG, MPFE, are used to estimate the qualities of the different algorithms. These experiments compare both of NSGA-Gs with Min point and Random metric to other algorithms.

Table 7. Generational Distance

	m	eMOEA	NSGA-R	NSGA-II	MOEA/D	NSGA-III	NSGA-G
DTLZ1	5	2.595e-01	4.418e-01	2.251e+01	4.264e-01	3.090e+00	1.977e-01
DTLZ3	5	1.861e-01	5.528e-02	1.130e+00	8.650e-02	3.079e-01	1.678e-02
WFG1	5	1.133e-03	9.748e-04	6.923e-03	6.908e-03	3.218e-03	7.617e-04
WFG3	5	4.027e-04	0.000e+00	2.549e-03	1.941e-03	2.011e-03	1.061e-05
DTLZ1	6	2.903e+00	2.137e+00	9.131e+01	1.820e+00	6.839e+00	4.907e-01
DTLZ3	6	2.226e+01	1.332e+01	1.252e+02	1.760e+01	2.389e+01	5.457e+00
WFG1	6	1.207e-03	8.842e-04	8.000e-03	6.753e-03	3.559e-03	7.417e-04
WFG3	6	4.104e-04	0.000e+00	2.523e-03	1.639e-03	1.800e-03	5.384e-05
DTLZ1	7	7.790e-01	8.949e-01	2.228e+01	2.601e-01	1.407e+00	8.201e-02
DTLZ3	7	1.719e-01	4.449e-02	1.309e+00	3.610e-02	1.619e-01	5.628e-03
WFG1	7	1.048e-03	8.219e-04	6.825e-03	5.613e-03	3.891e-03	6.405e-04
WFG3	7	4.011e-04	3.055e-06	2.390e-03	1.871e-03	1.665e-03	5.926e-05
DTLZ1	8	5.823e+00	5.851e+00	1.130e+02	1.276e+00	9.933e+00	4.660e-01
DTLZ3	8	2.071e+01	1.941e+01	1.604e+02	1.355e+01	3.001e+01	4.757e+00
WFG1	8	1.377e-03	9.406e-04	9.023e-03	7.659e-03	4.454e-03	6.469e-04
WFG3	8	3.655e-04	2.689e-05	1.692e-03	1.301e-03	9.662e-04	6.578e-05
DTLZ1	9	8.374e-01	3.626e+00	3.074e+01	3.544e-01	2.772e+00	1.003e-01
DTLZ3	9	4.673e-02	7.112e-02	6.293e-01	8.922e-03	1.052e-01	2.843e-03
WFG1	9	1.309e-03	8.924e-04	8.882e-03	7.551e-03	4.020e-03	6.816e-04
WFG3	9	3.597e-04	2.576e-05	1.298e-03	1.208e-03	7.634e-04	5.365e-05
DTLZ1	10	7.375e-01	1.519e+00	2.091e+01	2.705e-01	2.207e+00	3.021e-02
DTLZ3	10	4.785e-02	1.116e-01	6.793e-01	7.345e-03	1.118e-01	2.939e-03
WFG1	10	1.369e-03	1.385e-03	8.551e-03	6.364e-03	3.648e-03	6.692e-04
WFG3	10	3.259e-04	0.000e+00	1.196e-03	1.265e-03	6.945e-04	4.352e-05

Table 8. Average compute time in Generational Distance experiment

	m	eMOEA	NSGA-R	NSGA-II	MOEA/D	NSGA-III	NSGA-G
DTLZ1	5	3.604e+01	6.642e+01	5.508e+01	2.000e+02	2.241e+02	6.366e+01
DTLZ3	5	5.398e+01	6.440e+01	7.074e+01	1.870e+02	2.714e+02	6.212e+01
WFG1	5	1.379e+02	6.658e+01	6.636e+01	1.899e+02	2.594e+02	6.720e+01
WFG3	5	8.562e+02	8.162e+01	6.074e+01	1.864e+02	3.077e+02	8.370e+01
DTLZ1	6	4.552e+01	5.582e+01	5.632e+01	1.918e+02	1.662e+02	5.672e+01
DTLZ3	6	9.340e+01	6.572e+01	6.362e+01	1.971e+02	1.783e+02	6.638e+01
WFG1	6	1.961e+02	9.826e+01	7.392e+01	2.049e+02	2.157e+02	7.286e+01
WFG3	6	1.083e+03	7.580e+01	6.642e+01	1.967e+02	2.384e+02	7.782e+01
DTLZ1	7	6.206e+01	5.834e+01	6.208e+01	2.290e+02	1.621e+02	5.964e+01
DTLZ3	7	1.568e+02	6.992e+01	7.024e+01	2.405e+02	1.817e+02	7.022e+01
WFG1	7	2.585e+02	7.806e+01	8.042e+01	2.473e+02	2.085e+02	7.810e+01
WFG3	7	1.469e+03	8.030e+01	9.184e+01	2.896e+02	2.821e+02	9.950e+01
DTLZ1	8	8.762e+01	5.998e+01	6.640e+01	2.450e+02	2.327e+02	6.244e+01
DTLZ3	8	2.235e+02	7.618e+01	7.652e+01	2.536e+02	2.535e+02	7.424e+01
WFG1	8	3.100e+02	8.034e+01	8.710e+01	2.625e+02	2.924e+02	8.206e+01
WFG3	8	1.464e+03	7.912e+01	7.772e+01	2.542e+02	3.268e+02	8.346e+01
DTLZ1	9	1.157e+02	6.264e+01	7.034e+01	2.524e+02	3.095e+02	6.590e+01
DTLZ3	9	2.978e+02	7.694e+01	8.422e+01	2.678e+02	3.422e+02	7.828e+01
WFG1	9	3.846e+02	8.442e+01	9.426e+01	2.731e+02	3.844e+02	8.668e+01
WFG3	9	1.677e+03	8.954e+01	8.166e+01	2.595e+02	4.373e+02	8.642e+01
DTLZ1	10	1.527e+02	6.510e+01	7.584e+01	2.740e+02	4.204e+02	6.874e+01
DTLZ3	10	3.860e+02	8.132e+01	8.916e+01	2.883e+02	4.641e+02	8.370e+01
WFG1	10	4.747e+02	8.996e+01	1.005e+02	2.941e+02	5.175e+02	9.272e+01
WFG3	10	1.881e+03	8.576e+01	8.640e+01	2.802e+02	6.035e+02	9.128e+01

Table 9. Inverted Generational Distance

	m	eMOEA	NSGA-R	NSGA-II	MOEA/D	NSGA-III	NSGA-G
DTLZ1	5	3.437e-01	1.027e+00	3.741e+01	6.226e-01	3.465e+00	4.637e-01
DTLZ3	5	5.568e-01	4.794e-01	3.576e+00	3.969e-01	1.098e+00	1.589e-01
WFG1	5	1.298e-01	2.924e-01	1.234e-01	7.202e-02	1.365e-01	2.906e-01
WFG3	5	4.167e-02	3.850e-01	1.272e-01	1.417e-01	7.899e-02	3.987e-01
DTLZ1	6	4.975e+00	6.617e+00	2.469e+02	2.903e+00	9.524e+00	2.688e+00
DTLZ3	6	1.131e+02	4.698e+01	5.199e+02	4.207e+01	8.253e+01	2.761e+01
WFG1	6	1.722e-01	3.705e-01	1.531e-01	7.460e-02	1.596e-01	3.341e-01
WFG3	6	5.367e-02	5.424e-01	1.488e-01	1.630e-01	1.065e-01	5.146e-01
DTLZ1	7	7.034e-01	4.042e+00	1.938e+01	4.718e-01	7.695e-01	8.458e-01
DTLZ3	7	7.320e-01	4.310e-01	4.852e+00	2.878e-01	3.826e-01	2.524e-01
WFG1	7	1.437e-01	3.547e-01	1.371e-01	7.114e-02	1.403e-01	3.199e-01
WFG3	7	6.134e-02	6.325e-01	1.573e-01	1.705e-01	1.169e-01	6.122e-01
DTLZ1	8	1.234e+01	1.212e+01	4.166e+02	3.101e+00	1.073e+01	2.849e+00
DTLZ3	8	1.501e+02	6.557e+01	7.623e+02	3.720e+01	1.011e+02	2.665e+01
WFG1	8	1.284e-01	3.186e-01	1.251e-01	6.956e-02	1.238e-01	2.692e-01
WFG3	8	6.487e-02	6.477e-01	1.593e-01	1.704e-01	1.115e-01	6.094e-01
DTLZ1	9	4.009e-01	3.676e+00	5.490e+00	3.932e-01	6.185e-01	5.747e-01
DTLZ3	9	3.029e-01	4.578e-01	1.713e+00	2.398e-01	2.584e-01	2.401e-01
WFG1	9	1.167e-01	2.921e-01	1.193e-01	6.477e-02	1.131e-01	2.561e-01
WFG3	9	6.758e-02	6.897e-01	1.621e-01	1.675e-01	1.078e-01	6.237e-01
DTLZ1	10	9.350e-01	9.074e+00	1.357e+01	6.061e-01	1.499e+00	1.028e+00
DTLZ3	10	4.368e-01	5.440e-01	2.368e+00	2.000e-01	3.965e-01	1.912e-01
WFG1	10	1.147e-01	3.043e-01	1.167e-01	6.273e-02	1.102e-01	2.671e-01
WFG3	10	6.759e-02	6.676e-01	1.670e-01	1.696e-01	1.043e-01	6.102e-01

Table 10. Average compute time in Inverted Generational Distance experiment

	m	eMOEA	NSGA-R	NSGA-II	MOEA/D	NSGA-III	NSGA-G
DTLZ1	5	3.384e+01	5.500e+01	5.176e+01	1.840e+02	2.139e+02	5.276e+01
DTLZ3	5	9.490e+01	8.072e+01	5.954e+01	2.942e+02	2.803e+02	6.146e+01
WFG1	5	1.453e+02	7.752e+01	8.710e+01	1.957e+02	2.988e+02	8.220e+01
WFG3	5	9.067e+02	8.638e+01	5.950e+01	2.087e+02	3.137e+02	8.416e+01
DTLZ1	6	4.982e+01	6.264e+01	5.860e+01	2.209e+02	1.894e+02	6.534e+01
DTLZ3	6	9.604e+01	6.984e+01	6.554e+01	2.182e+02	1.958e+02	7.078e+01
WFG1	6	2.188e+02	8.088e+01	7.810e+01	2.452e+02	2.282e+02	8.362e+01
WFG3	6	2.601e+03	9.036e+01	6.638e+01	3.200e+02	3.094e+02	1.215e+02
DTLZ1	7	6.754e+01	5.880e+01	6.122e+01	2.517e+02	1.620e+02	6.066e+01
DTLZ3	7	1.587e+02	7.172e+01	6.986e+01	2.525e+02	1.798e+02	7.168e+01
WFG1	7	2.579e+02	7.696e+01	8.294e+01	2.587e+02	2.185e+02	7.768e+01
WFG3	7	1.272e+03	7.836e+01	7.194e+01	2.487e+02	2.284e+02	8.888e+01
DTLZ1	8	8.430e+01	5.996e+01	6.610e+01	2.537e+02	2.322e+02	6.328e+01
DTLZ3	8	2.358e+02	7.418e+01	7.808e+01	2.608e+02	2.535e+02	7.446e+01
WFG1	8	3.158e+02	7.960e+01	8.682e+01	2.704e+02	2.903e+02	8.344e+01
WFG3	8	1.432e+03	8.044e+01	7.712e+01	2.513e+02	3.242e+02	8.364e+01
DTLZ1	9	1.237e+02	6.278e+01	6.978e+01	2.563e+02	3.120e+02	6.646e+01
DTLZ3	9	3.174e+02	7.882e+01	8.330e+01	2.721e+02	3.418e+02	7.838e+01
WFG1	9	3.827e+02	8.586e+01	9.338e+01	2.718e+02	3.837e+02	8.594e+01
WFG3	9	1.696e+03	8.290e+01	8.142e+01	2.607e+02	4.369e+02	8.654e+01
DTLZ1	10	1.436e+02	6.472e+01	7.536e+01	2.753e+02	4.187e+02	6.876e+01
DTLZ3	10	4.003e+02	8.566e+01	8.872e+01	2.897e+02	4.572e+02	8.270e+01
WFG1	10	4.635e+02	8.924e+01	1.008e+02	2.915e+02	5.137e+02	9.116e+01
WFG3	10	1.902e+03	8.662e+01	8.612e+01	2.802e+02	6.022e+02	9.028e+01

Table 11. Maximum Pareto Front Error

	m	eMOEA	NSGA-R	NSGA-II	MOEA/D	NSGA-III	NSGA-G
DTLZ1	5	2.008e+01	1.195e+01	8.083e+02	1.765e+01	3.548e+02	4.912e+00
DTLZ3	5	1.079e+01	1.564e+00	2.545e+01	1.604e+00	1.546e+01	5.798e-01
WFG1	5	1.332e-01	1.763e-02	2.620e-01	2.042e-01	1.709e-01	1.588e-02
WFG3	5	1.583e-01	0.000e+00	9.601e-02	6.763e-02	1.139e-01	0.000e+00
DTLZ1	6	2.937e+02	5.789e+01	1.583e+03	5.168e+01	3.920e+02	7.665e+00
DTLZ3	6	1.045e+03	2.861e+02	1.825e+03	1.913e+02	7.048e+02	7.409e+01
WFG1	6	2.288e-01	1.619e-02	3.790e-01	3.086e-01	2.649e-01	1.372e-02
WFG3	6	1.690e-01	0.000e+00	1.090e-01	7.179e-02	9.973e-02	2.058e-03
DTLZ1	7	1.193e+02	4.205e+01	8.990e+02	9.095e+00	1.081e+02	2.998e+00
DTLZ3	7	1.138e+02	2.539e+00	1.768e+01	3.267e+00	4.447e+00	1.286e-01
WFG1	7	2.461e-01	1.443e-02	3.670e-01	2.775e-01	2.428e-01	1.545e-02
WFG3	7	1.630e-01	6.556e-04	1.017e-01	6.336e-02	7.499e-02	2.411e-03
DTLZ1	8	4.798e+02	2.375e+02	1.982e+03	4.991e+01	5.619e+02	8.178e+00
DTLZ3	8	1.458e+03	3.881e+02	2.152e+03	1.856e+02	9.085e+02	6.259e+01
WFG1	8	2.722e-01	1.486e-02	4.113e-01	3.155e-01	3.020e-01	1.039e-02
WFG3	8	1.499e-01	0.000e+00	9.124e-02	5.919e-02	6.697e-02	2.380e-03
DTLZ1	9	1.732e+02	1.234e+02	9.926e+02	1.264e+01	3.271e+02	1.976e+00
DTLZ3	9	7.820e+00	3.242e+00	1.899e+01	2.121e-01	6.489e+00	9.978e-02
WFG1	9	2.388e-01	1.108e-02	3.644e-01	2.316e-01	2.435e-01	7.929e-03
WFG3	9	1.516e-01	4.995e-04	8.803e-02	5.787e-02	8.046e-02	1.736e-03
DTLZ1	10	1.097e+02	1.138e+02	9.838e+02	8.148e+00	3.040e+02	2.231e+00
DTLZ3	10	6.727e+00	2.405e+00	1.556e+01	1.584e-01	5.933e+00	7.632e-02
WFG1	10	3.030e-01	1.372e-02	4.268e-01	2.544e-01	3.118e-01	9.250e-03
WFG3	10	1.468e-01	3.964e-04	7.328e-02	5.557e-02	6.889e-02	2.378e-03

Table 12. Average compute time in Maximum Pareto Front Error experiment

	m	eMOEA	NSGA-R	NSGA-II	MOEA/D	NSGA-III	NSGA-G
DTLZ1	5	4.128e+01	5.408e+01	5.408e+01	2.401e+02	2.308e+02	5.522e+01
DTLZ3	5	5.676e+01	6.470e+01	5.944e+01	2.074e+02	2.982e+02	6.294e+01
WFG1	5	1.623e+02	8.048e+01	7.232e+01	2.239e+02	2.815e+02	7.082e+01
WFG3	5	9.397e+02	7.952e+01	6.154e+01	2.043e+02	3.174e+02	1.023e+02
DTLZ1	6	4.550e+01	5.556e+01	5.634e+01	1.924e+02	1.662e+02	5.686e+01
DTLZ3	6	9.168e+01	6.554e+01	6.418e+01	1.985e+02	1.787e+02	6.656e+01
WFG1	6	1.958e+02	7.512e+01	7.434e+01	2.072e+02	2.170e+02	7.650e+01
WFG3	6	1.136e+03	7.774e+01	6.724e+01	1.967e+02	2.406e+02	7.916e+01
DTLZ1	7	8.734e+01	6.188e+01	6.164e+01	2.453e+02	2.058e+02	6.204e+01
DTLZ3	7	1.622e+02	7.046e+01	7.170e+01	2.699e+02	1.812e+02	8.968e+01
WFG1	7	2.674e+02	8.156e+01	8.470e+01	2.574e+02	2.154e+02	8.036e+01
WFG3	7	1.461e+03	8.358e+01	7.426e+01	2.546e+02	2.334e+02	8.180e+01
DTLZ1	8	8.920e+01	6.054e+01	6.592e+01	2.443e+02	2.349e+02	6.252e+01
DTLZ3	8	2.360e+02	7.318e+01	7.644e+01	2.536e+02	2.555e+02	7.426e+01
WFG1	8	4.476e+02	7.960e+01	8.678e+01	2.612e+02	2.932e+02	8.164e+01
WFG3	8	1.482e+03	8.250e+01	7.690e+01	2.497e+02	3.244e+02	8.380e+01
DTLZ1	9	1.031e+02	6.208e+01	6.984e+01	2.514e+02	3.068e+02	6.554e+01
DTLZ3	9	3.043e+02	7.924e+01	8.222e+01	2.634e+02	3.368e+02	7.806e+01
WFG1	9	3.935e+02	8.856e+01	9.290e+01	2.700e+02	3.807e+02	8.676e+01
WFG3	9	1.660e+03	9.016e+01	8.028e+01	2.594e+02	4.347e+02	8.622e+01
DTLZ1	10	1.507e+02	6.436e+01	7.442e+01	2.728e+02	4.151e+02	6.830e+01
DTLZ3	10	3.933e+02	8.494e+01	8.852e+01	2.865e+02	4.593e+02	8.244e+01
WFG1	10	4.769e+02	9.296e+01	9.974e+01	2.904e+02	5.110e+02	9.182e+01
WFG3	10	1.875e+03	8.474e+01	8.594e+01	2.784e+02	6.013e+02	9.096e+01

First, two versions of NSGA-G often show that they are faster than the other algorithms in all experiments of average computation time, Table 8, 10, and 12.

Second, NSGA-Gs are also better than other NSGAs in terms of quality in GD and MPFE experiments, as shown in Table 7, and 11. Except for the IDG experiment, as shown in Table 9 the quality of NSGA-G with Random metric is not as good as other ones. However, the fastest algorithm among NSGAs is often NSGA-G with random metric. It can be accepted for the trade-off between quality and computation time.

Study on the Evaluation. In the previous experiments, we survey algorithms with various problems and the constant number of max evaluation. This section selects a specific problem and shows the observation of algorithms while the process is running. In particular, we choose DTLZ3 problem with eight objectives, called DTLZ3-8. Besides, we focus on reducing the execution time of NSGAs algorithm. Hence, this section compares two versions of NSGA-G algorithms to others in NSGA class, such as NSGA-II and NSGA-III. Two versions of NSGA-G with Min point and Random metric are called NSGA-G and NSGA-R, respectively.

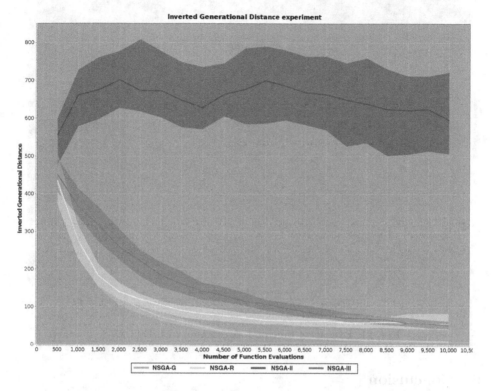

Fig. 10. Inverted Generational Distance of 4 algorithms with DTLZ3-8.

The results in Fig. 10 and 11 show that two versions of NSGA-G are faster than others. Both their convergence and diversity are better than NSGA-II and NSGA-III.

In conclusion, NSGA-Gs often show better quality and faster execution time in most cases, such as DTLZs, WFGs. One main conclusion of these experiments is that NSGA-G with a Random metric is often the least expensive in terms of computation.

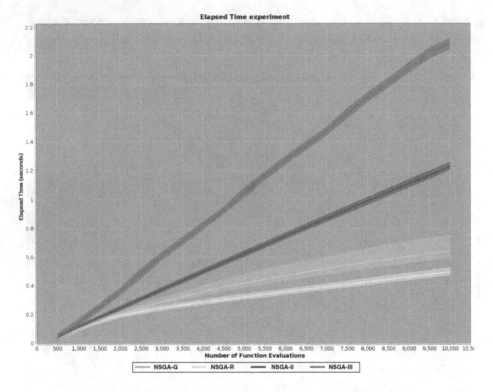

Fig. 11. Execution time of 4 algorithms with DTLZ3-8.

6 Conclusion

This paper is about medical data management in cloud federation. It introduces Dynamic Regression Algorithm (DREAM) as a part of MIDAS and on top of IReS, an open source platform for complex analytics work-flows executed over multi-engine environments. DREAM aims to address variance in a cloud federation and to provide accurate estimation for MOQP. Experiment results with DREAM and TPC-H benchmark are quite promising with respect to existing solutions. Further more, we introduce Non-dominated Sorting Algorithms based on Grid partitioning (NSGA-G) in searching and optimization MOOP. We validated NSGA-Gs with DTLZ, WFG test problems, and MOEA framework. The experiments show that NSGA-Gs often show better quality and faster execution time than other NSGAs in most cases, such as DTLZs, WFGs. One main conclusion of these experiments is that NSGA-G with a Random metric is often the least expensive in terms of computation.

In the future, we plan to validate our proposal with more cloud providers (and their associated pricing model and services) and data management systems. We will also define new strategies to choose QEPs in a Pareto Set. Further more, the size of population in each generation iterate is constant in many NSGAs.

The suitable value of population size is still a question of NSGAs. Future works include a deeper study on the impact of the size of the population.

References

1. Abadi, D., et al.: The Beckman report on database research. J. Commun. ACM **59**(2), 92–99 (2016)
2. Akdere, M., Çetintemel, U., Riondato, M., Upfal, E., Zdonik, S.B.: Learning-based query performance modeling and prediction. In: 2012 IEEE 28th International Conference on Data Engineering, Washington, DC, pp. 390–401 (2012)
3. Armbrust, M., et al.: A view of cloud computing. Commun. ACM **53**(4), 50–58 (2010)
4. Bezerra, L.C.T., López-Ibáñez, M., Stützle, T.: An empirical assessment of the properties of inverted generational distance on multi- and many-objective optimization. In: Trautmann, H., Rudolph, G., Klamroth, K., Schütze, O., Wiecek, M., Jin, Y., Grimme, C. (eds.) EMO 2017. LNCS, vol. 10173, pp. 31–45. Springer, Cham (2017). https://doi.org/10.1007/978-3-319-54157-0_3
5. Breiman, L.: Bagging predictors. Mach. Learn. **24**(2), 123–140 (1996)
6. Bugiotti, F., Bursztyn, D., Deutsch, A., Ileana, I., Manolescu, I.: Invisible glue: scalable self-tuning multi-stores. In: Conference on Innovative Data Systems Research (CIDR), Asilomar, CA, USA (2015)
7. Cerf, R.: Asymptotic convergence of genetic algorithms. In: Advances in Applied Probability, vol. 30, no. 2, pp. 521–550. Cambridge University Press (1998)
8. Chankong, V., Haimes, Y.Y.: Multiobjective Decision Making: Theory and Methodology, North-Holland Series in System Science and Engineering, North Holland (1983)
9. Coello, C.A.C., Cortés, N.C.: Solving multiobjective optimization problems using an artificial immune system. Genet. Program. Evolvable Mach. **6**, 163–190 (2005)
10. Coello, C.A.C., Lamont, G.B., Veldhuizen, D.A.V.: Evolutionary Algorithms for Solving Multi-objective Problems. Genetic and Evolutionary Computation, 2nd edn., pp. I-XXI, 1-800. Springer, New York (2007). https://doi.org/10.1007/978-0-387-36797-2. ISBN 978-0-387-33254-3
11. DeWitt, D.J., et al.: Split query processing in polybase. In: Proceedings of the 2013 ACM SIGMOD International Conference on Management of Data (SIGMOD 2013), pp. 1255–1266. Association for Computing Machinery, New York (2013)
12. Dean, J., Ghemawat, S.: MapReduce: simplified data processing on large clusters. Commun. ACM **51**(1), 107–113 (2008)
13. Deb, K., Agrawal, R.B.: Simulated binary crossover for continuous search space. Complex Syst. **9**, 1–34 (1994)
14. Deb, K., Jain, H.: An evolutionary many-objective optimization algorithm using reference-point-based nondominated sorting approach, Part I: solving problems with box constraints. IEEE Trans. Evol. Comput. **18**(4), 577–601 (2014)
15. Deb, K., Pratap, A., Agarwal, S., Meyarivan, T.: A fast and elitist multiobjective genetic algorithm: NSGA-II. IEEE Trans. Evol. Comput. **6**(2), 182–197 (2002)
16. Deb, K., Thiele, L., Laumanns, M., Zitzler, E.: Scalable test problems for evolutionary multiobjective optimization. In: Abraham, A., Jain, L., Goldberg, R. (eds.) Evolutionary Multiobjective Optimization. Advanced Information and Knowledge Processing, pp. 105–145. Springer, London (2005). https://doi.org/10.1007/1-84628-137-7_6

17. Doka, K., Papailiou, N., Tsoumakos, D., Mantas, C., Koziris, N.: IReS: intelligent, multi-engine resource scheduler for big data analytics workflows. In: Proceedings of the: ACM SIGMOD International Conference on Management of Data (SIGMOD 2015), pp. 1451–1456. ACM, New York (2015)
18. Elmore, A., et al.: A demonstration of the BigDAWG polystore system. Proc. VLDB Endow. **8**(12), 1908–1911 (2015)
19. Fard, H.M., Prodan, R., Barrionuevo, J.J.D., Fahringer, T.: A multi-objective approach for workflow scheduling in heterogeneous environments. In: 12th IEEE/ACM International Symposium on Cluster, Cloud and Grid Computing (ccgrid 2012), Ottawa, ON, pp. 300–309 (2012)
20. Fonseca, C.M., Fleming, P.J.: An overview of evolutionary algorithms in multiobjective optimization. Evol. Comput. **3**(1), 1–16 (1995)
21. Ganapathi, A., et al.: Predicting multiple metrics for queries: better decisions enabled by machine learning. In: 2009 IEEE 25th International Conference on Data Engineering, Shanghai, pp. 592–603(2009)
22. Giannakouris, V., Papailiou, N., Tsoumakos, D., Koziris, N.: MuSQLE: distributed SQL query execution over multiple engine environments. In: 2016 IEEE International Conference on Big Data (Big Data), Washington, DC, pp. 452–461 (2016)
23. Glaßer, C., Reitwießner, C., Schmitz, H., Witek, M.: Approximability and hardness in multi-objective optimization. In: Ferreira, F., Löwe, B., Mayordomo, E., Mendes Gomes, L. (eds.) CiE 2010. LNCS, vol. 6158, pp. 180–189. Springer, Heidelberg (2010). https://doi.org/10.1007/978-3-642-13962-8_20
24. Helff, F., Gruenwald, L., D'Orazio, L.: Weighted sum model for multi-objective query optimization for mobile-cloud database environments. In: EDBT/ICDT Workshops (2016)
25. Huband, S., Hingston, P., Barone, L., While, L.: A review of multiobjective test problems and a scalable test problem toolkit. IEEE Trans. Evol. Comput. **10**(5), 447–506 (2006)
26. Ishibuchi, H., Masuda, H., Nojima, Y.: Sensitivity of performance evaluation results by inverted generational distance to reference points. In: IEEE Congress on Evolutionary Computation (CEC), pp. 1107–1114, July 2016
27. Jain, H., Deb, K.: An evolutionary many-objective optimization algorithm using reference-point based nondominated sorting approach, Part II: handling constraints and extending to an adaptive approach. IEEE Trans. Evol. Comput. **18**(4), 602–622 (2014)
28. Karpathiotakis, M., Alagiannis, I., Ailamaki, A.: Fast queries over heterogeneous data through engine customization. Proc. VLDB Endow. **9**(12), 972–983 (2016)
29. Khan, S.A., Rehman, S.: Iterative non-deterministic algorithms in on-shore wind farm design: a brief survey. Renew. Sustain. Energy Rev. **19**, 370–384 (2013)
30. Kllapi, H., Sitaridi, E., Tsangaris, M.M., Ioannidis, Y.: Schedule optimization for data processing flows on the cloud. In: Proceedings of the 2011 International Conference on Management of Data - SIGMOD 2011, pp. 289 (2011)
31. Knowles, J., Corne, D.: The Pareto archived evolution strategy: a new baseline algorithm for Pareto multiobjective optimisation. In: Proceedings of the 1999 Congress on Evolutionary Computation-CEC99 (Cat. No. 99TH8406), Washington, DC, USA, vol. 1, pp. 98–105 (1999)
32. Kolev, B., Bondiombouy, C., Valduriez, P., Jimenez-Peris, R., Pau, R., Pereira, J.: The CloudMdsQL multistore system. In: Proceedings of the 2016 International Conference on Management of Data (SIGMOD 2016), pp. 2113–2116. Association for Computing Machinery, New York (2016)

33. Kolev, B., Valduriez, P., Bondiombouy, C., Jiménez-Peris, R., Pau, R., Pereira, J.: CloudMdsQL: querying heterogeneous cloud data stores with a common language. Distrib. Parallel Database **34**(4), 463–503 (2016)
34. Köppen, M., Yoshida, K.: Substitute distance assignments in NSGA-II for handling many-objective optimization problems. In: Obayashi, S., Deb, K., Poloni, C., Hiroyasu, T., Murata, T. (eds.) EMO 2007. LNCS, vol. 4403, pp. 727–741. Springer, Heidelberg (2007). https://doi.org/10.1007/978-3-540-70928-2_55
35. Kurze, T., Klems, M., Bermbach, D., Lenk, A., Tai, S., Kunze, M.: Cloud federation. In: The Second International Conference on Cloud Computing, GRIDs, and Virtualization, pp. 32–38 (2011)
36. Le, T.-D., Kantere, V., D'Orazio, L.: An efficient multi-objective genetic algorithm for cloud computing: NSGA-G. In: IEEE International Conference on Big Data, Big Data 2018, Seattle, WA, USA, 10–13 December, pp. 3883–3888 (2018)
37. Le, T.-D., Kantere, V., D'Orazio, L.: Dynamic estimation for medical data management in a cloud federation. In: Proceedings of the Workshops of the EDBT/ICDT 2019 Joint Conference, EDBT/ICDT 2019, Lisbon, Portugal, 26 March 2019 (2019)
38. LeFevre, J., Sankaranarayanan, J., Hacigümüs, H., Tatemura, J., Polyzotis, N., Carey, M.J.: MISO: souping up big data query processing with a multistore system. In: SIGMOD Conference 2014 , pp. 1591–1602 (2014)
39. Nykiel, T., Potamias, M., Mishra, C., Kollios, G., Koudas, N.: MRShare: sharing across multiple queries in MapReduce. PVLDB **3**, 494–505 (2010)
40. Papakonstantinou, Y.: Polystore query rewriting: the challenges of variety. In: EDBT/ICDT Workshops (2016)
41. Rousseeuw, P.J., Leroy, A.M.: Robust Regression and Outlier Detection. Wiley, New York (1987)
42. Sidhanta, S., Golab, W., Mukhopadhyay, S.: OptEx: a deadline-aware cost optimization model for spark. In: 16th IEEE/ACM International Symposium on Cluster, Cloud and Grid Computing (CCGrid), Cartagena, pp. 193–202 (2016)
43. Soong, T.T.: Fundamentals of Probability and Statistics for Engineers. Wiley, New York (2004)
44. Srinivas, N., Deb, K.: Muiltiobjective optimization using nondominated sorting in genetic algorithms. Evol. Comput. **2**(3), 221–248 (1994)
45. Thusoo, A., et al.: Hive - a petabyte scale data warehouse using Hadoop. In: 2010 IEEE 26th International Conference on Data Engineering (ICDE 2010), pp. 996–1005 (2010)
46. Tozer, S., Brecht, T., Aboulnaga, A.: Q-Cop: avoiding bad query mixes to minimize client timeouts under heavy loads. In: 2010 IEEE 26th International Conference on Data Engineering (ICDE 2010), Long Beach, CA, pp. 397–408 (2010)
47. Trummer, L., Koch, C.: Multi-objective parametric query optimization. Commun. ACM **60**(10), 81–89 (2017)
48. Veldhuizen, D.A.V.: Multiobjective evolutionary algorithms: classifications, analyses, and new innovations. Ph.D. thesis, Department of Electrical and Computer Engineering. Graduate School of Engineering. Air Force Institute of Technology, Wright-Patterson AFB, Ohio (1999)
49. Veldhuizen, D.A.V., Lamont, G.B.: Evolutionary computation and convergence to a pareto front. In: Late Breaking Papers at the Genetic Programming 1998 Conference, pp. 221–228 (1998)
50. Wu, W., Chi, Y., Zhu, S., Tatemura, J., Hacigümüs, H., Naughton, J.F.: Predicting query execution time: are optimizer cost models really unusable? In: 2013 IEEE 29th International Conference on Data Engineering (ICDE), Brisbane, QLD, pp. 1081–1092 (2013)

51. Xiong, P., Chi, Y., Zhu, S., Tatemura, J., Pu, C., Hacigümüş, H.: ActiveSLA: a profit-oriented admission control framework for Database-as-a-Service providers. In: Proceedings of the 2nd ACM Symposium on Cloud Computing, SOCC 2011 (2011)
52. Yen, G.G., He, Z.: Performance metrics ensemble for multiobjective evolutionary algorithms. IEEE Trans. Evol. Comput. **18**, 131–144 (2013)
53. Zhang, Q., Li, H.: MOEA/D: a multiobjective evolutionary algorithm based on decomposition. IEEE Trans. Evol. Comput. **11**(6), 712–731 (2007)
54. Zitzler, E., Laumanns, M., Thiele, L.: SPEA2: improving the strength pareto evolutionary algorithm, TIK-Report. 103 (2001)
55. Zitzler, E., Thiele, L., Laumanns, M., Fonseca, C.M., Fonseca, V.G.D.: Performance assessment of multiobjective optimizers: an analysis and review. IEEE Trans. Evol. Comput. **7**(2), 117–132 (2003)

Temporal Pattern Mining
for E-commerce Dataset

Mohamad Kanaan[1]($\boxtimes$), Remy Cazabet[2], and Hamamache Kheddouci[2]

[1] Sistema-Strategy, Lyon, France
mohamad.kanaan@sistema-strategy.com
[2] Claude Bernard University Lyon 1, LIRIS Lab, Lyon, France
{remy.cazabet,hamamache.kheddouci}@univ-lyon1.fr

Abstract. Over the last few years, several data mining algorithms have been developed to understand customers' behaviors in e-commerce platforms. They aim to extract knowledge and predict future actions on the website. In this paper we present three algorithms: **SEPM−**, **SEPM+** and **SEPM++** (Sequential Event Pattern Mining), for mining sequential frequent patterns. Our goal is to mine clickstream data to extract and analyze useful sequential patterns of clicks. For this purpose, we augment the vertical representation of patterns with additional information about the items' duration. Then based on this representation, we propose the necessary algorithms to mine sequential frequent patterns with the average duration of each of their items. Also, the direction of durations' variation in the sequence is taken into account by the algorithms. This duration is used as a proxy of the interest of the user in the content of the page. Finally, we categorize the resulting patterns and we prove that they are more discriminating than the standard ones. Our approach is tested on real data, and patterns found are analyzed to extract users' discriminatory behaviors. The experimental results on both real and synthetic datasets indicate that our algorithms are efficient and scalable.

Keywords: E-commerce · Customer behavior · Data mining · Sequential frequent pattern

1 Introduction

Electronic (e)-commerce can be defined as the use of electronic media for carrying out commercial transactions to exchange goods or services. It has revolutionized retail by reducing the distance between stores and customers. Nowadays, customers' choices are no longer limited to products proposed in their regions. They can purchase from any international store, anywhere and anytime, by using an e-commerce website. This attractivity has significantly increased the e-commerce web sites' visitors, motivating e-sellers to improve their websites and to propose more personalized products, by better understanding the behaviors of their customers. These behaviors are very complex since they are influenced by various factors, e.g., the customers' profile (age, gender, country, job, ...), the services

© Springer-Verlag GmbH Germany, part of Springer Nature 2020
A. Hameurlain and A M. Tjoa (Eds.): TLDKS XLVI, LNCS 12410, pp. 67–90, 2020.
https://doi.org/10.1007/978-3-662-62386-2_3

proposed (payment method, product diversity, delivery method, ...), and so on. Understanding these behaviors help e-sellers to recommend relevant products, predict future purchases, and ensure products' availability. To further improve their website performances, e-sellers want to know how customers purchase products, how they navigate through the product catalogs, or even why they abandon some purchase processes.

To achieve this goal, many tools have been developed to trace the users' actions (click, view, add to cart, purchase, etc.). These traces are commonly known as **clickstreams**. Then, by using some data mining techniques, clickstreams are mined to extract useful knowledge. Some well-known data mining techniques used for this purpose are: pattern mining [15,25], trend discovery [22,38], customers clustering and classification [2,6,11], collaborative filtering in recommending systems [28,29,37] and purchase prediction [20,26,41].

In this paper, our goal is to discover how the users navigate through a website to find their products. We, therefore, extract frequent discriminatory behaviors, and we analyze their skeletons.

These frequent discriminatory behaviors can be present in clickstreams as sequential frequent patterns. A sequential pattern is an ordered, common, and frequent sequence of actions (items) that lead to a specific action (e.g., purchase or abandon). An example of sequential pattern can be: {if a customer **buy** a *printer* → he is likely to **buy** a *cartridge* right after}.

In previous years, several sequential pattern mining algorithms [15,25] have been proposed. They are designed to mine only inputs containing the labels of items. However, clickstreams do not contain only items labels, but also contain the items' duration. This duration represents the time spent by a user on a page, and it can be used as a proxy of user interest in the content of the page. A user could be more interested in a product that he has viewed for 4 min than another product that he has viewed for 5 s.

We decide to take this duration into account and integrate it into the mining process, to make the extracted patterns more significant. Therefore, two aspects are taken into consideration: the order of browsed products and the time spent by customers on consulted pages. The order is used to keep the sequential nature of data, and the duration to make the patterns more meaningful. The objective is to find the **sequence** of products consulted **frequently** with the **average duration** time spent on each product.

This paper is organized as follows. Section 2 presents a state of the art of the frequent pattern mining literature. The problem is reformulated in Sect. 3 and in Sect. 4 a new algorithm is developed to mine sequential event pattern. The experimental results are shown in Sect. 5, before concluding in Sect. 6.

1.1 Motivations

E-sellers know that their customers are all different, but they also know that there are regularities in their product searching. Understanding these subtleties can help them to personalize their website according to their customers' profiles, and propose new commercial offers.

In this paper, our main goal is to provide e-sellers with appropriate techniques to acquire knowledge of their data by designing the necessary tools to extract their customers' behaviors. For example, some of these discriminatory behaviors can be:

1. *classic behavior*: customers consult the same sequence of products and spent pretty much the same time on them before purchasing a product
2. *observer behavior*: customers, before focusing on their main products, consult some related "accessories" (e.g., mobile phone accessories before mobile phones, printer cartridges before printers, ...)
3. *comparator behavior*: customers consult several times the same products to compare the properties of each of them
4. *research behavior:* customers check the product catalog randomly without purchase intention

These behaviors can be modeled as sequential frequent patterns that contain a sequence of products consulted in the same order several times. Many algorithms are proposed to mine the sequential frequent patterns where patterns are represented by only a sequence of the label. In these algorithms, the duration is not used during mining patterns. However, this duration can be very helpful for the analysis of usages. It is an important factor that can be used to reflect discriminatory behaviors. In our work, we choose to integrate the duration into these patterns. It will make them more talkative and allow an easier matching between patterns and their discriminatory behaviors than the classic ones.

1.2 Contribution

This work is an extension of our previous work [21], in which we proposed SEPM (Sequential Event Pattern Mining). SEPM is able to extract the sequential patterns with the average duration of their items. Nevertheless, an important temporal factor was not taken into account: the direction of variation of this duration. In addition to the work done in [21], we show here how the variation of duration is integrated into the representation of patterns, and into the mining process, and how this variation may help to detect more discriminatory patterns. Also, we detail the algorithms and we propose a multi-threaded version.

So, in this paper, our goal remains the same with a difference in how patterns are modeled and mined. They become more discriminatory. We pay more attention to the variation of their duration during the mining process to preserve the shape of the temporal sequence. Our tests show that taking this variation into account can help to obtain better results and more interpretable patterns.

2 Related Work

The **sequence mining** task was proposed for the first time in [1] as a problem of mining customers' sales transactions in order to discover frequent sequences of purchase constrained by a user-specified minimum support. Later on, it has been used in several fields of application: medical treatments [5,18,35], webpage clickstream [19,27,31,34,40], telephone calls [42], DNA sequences [23,36], and so on and so forth. These different applications motivated researchers in the field to improve existing algorithms (e.g., on execution time, memory consumption, ...) and to propose related methods able to answer a slightly different question, highlighting different aspects of the data (closed sequential patterns [12,17], maximal sequential patterns [13,16,24], ...). In our work, we are interested in the sequential pattern mining problem.

Numerous sequence mining algorithms have been developed (e.g., [15,25]) as extensions of the sequential pattern problem, to solve problems specific to an application case. All these sequential patterns mining algorithms have the same input and output but the difference is in their mining technique. Their common goal is to find all possible and valid sequential frequent patterns. Given a set of sequences and a support threshold *minSupp* (user-specified minimum support), the goal is to find the subsequences that appear at least in *minSupp* sequences. One of the first proposed algorithms is **GSP** [33], which uses a breadth-first search to explore the candidate space and extract the patterns level by level. It is based on the AprioriAll algorithm [1]. It starts by scanning the input database to extract all frequent sequences containing one frequent item called 1−pattern. Then, using these patterns, it generates those containing two frequent items ("the 2−patterns"), and so on until no new pattern is found. (k)-patterns found at level k are used to generate the $(k+1)$-patterns in the next level $(k+1)$. The GSP algorithm has two main objectives: candidate generation and support counting. At level $(k+1)$, GSP starts by generating all $(k+1)$-candidates (potential pattern) from the (k)-patterns. They are called candidates because we do not know yet their supports and therefore we do not know if they respect the frequency constraint or not. After the candidate generation step, the support counting task is performed, in which all candidates' supports are calculated. Those who have supported greater than or equal to *minSupp* become patterns and the others are rejected. These two tasks are repeated at each level until no new pattern is found. During the support counting task, GSP makes multiple passes over the database in order to verify and count the appearances of patterns in all sequences. This task may take a long execution time (caused by the multiple passes) and may consume a lot of memory as we need to keep the original database in memory.

To overcome this drawback, in [39] a vertical format to represent the patterns has been used. A new algorithm called SPADE is also proposed, which can be used in BFS (breadth-first search) or DFS (depth-first search) to mine sequential patterns. The patterns' positions in the original database are saved in a vertical table called IDList. This table contains the id of sequences and the positions in these sequences where the pattern appears. SPADE executes two main tasks at each level of pattern exploring: candidate generation and support counting. To generate a new candidate, two patterns are joined if they have the same sequence of items except the last one. For example, $p_1 = \{A, B, C\}$ can join $p_2 = \{A, B, D\}$ and this join operation will produce the following candidates:

$c_1 = \{A, B, C, D\}$, $c_2 = \{A, B, D, C\}$, and $c_3 = \{A, B, CD\}$. Finally, SPADE computes the support of all candidates and keep those who respect the frequency constraint. It continues to repeat the generation of new candidates until no new pattern is found. One of SPADE's strong points lies in the computing of the patterns' supports. This operation can be accomplished just by checking the vertical table of the pattern. As we mentioned before, this table contains the id of the sequences where the pattern is. Therefore, the support is equal to the number of distinct sequence IDs. Unlike GSP, SPADE does not need to re-scan the original database anymore for that reason. The architecture of our proposed algorithm is inspired by these algorithms.

Other algorithms such as PrefixSpan [32] and SPAM [7] are proposed later. PrefixSpan is a pattern-growth algorithm in which the sequence database is recursively projected into a set of smaller projected sequence databases and each one of them will grow by exploring locally frequent fragments. PrefixSpan starts by generating a search tree starting from an empty root node. Each child node in the search tree extends its parent node's sequence with items in addition then PrefixSpan explores the tree search space using a depth-first search. A total order on the item is used to never generate duplicate sequences in the search tree. SPAM represents sequences by using a vertical bit-vectors. It also explores the search space in a depth-first strategy with some pruning mechanisms. It is efficient especially with very long sequences.

For a few years, many temporal sequential patterns algorithms are proposed. One of those is IEMiner [30], that mines frequent temporal patterns from interval-based events. Each event in the discovered patterns has a temporal relation (defined by Allen's interval algebra [4]) with all the others. Two sequences of events are equal only if they have the same sequence of labels and the same sequence of temporal relations. The duration is not a determining factor in the proposed mining process. Other works [9, 21] take the duration into consideration, but without paying attention to its variation in a sequence.

We believe that the sequence of durations is also important to represent a pattern, as it gives the pattern more information (by adding the duration as a new dimension into the pattern representation) and allows the detection of more discriminating ones.

3   Problem Reformulation

As previously described, our goal is to mine the discriminatory frequent behaviors of users on e-commerce websites. These behaviors can be found in the form of **sequential patterns with duration** in the clickstream. A sequential pattern is composed of a frequent sequence of consulted items that may represent a specific frequent behavior (purchase, abandon, random navigation, ...). For example, a sequential pattern can be composed of several items consulted in the same order *{Samsung note 5 → Samsung note 3 → Samsung note 4}* before purchasing. The consultation time spent on each of these items is crucial since it can indicate whether the user is susceptible to being interested in the page (product) or not.

Thereby, we propose to include this duration into the mining process. Patterns will, therefore, consist of the sequence of items' labels and of the average time spent on those items. Following [39], we adopt a vertical representation of the dataset and patterns, allowing a more efficient discovery process.

Definition 1. *(event). An event E is represented by a couple (label, duration), where label is the event label and duration is its duration.*

Definition 2. *(event list). A list of sequential events $SE = \{E_1, E_2, ..., E_n\}$ is a list of events sorted by their starting time. A canonical representation of SE can be obtained by merging all its labels in the order of their appearance in SE (e.g., in Table 1, SE with SID = 3 can be represented by $<ABDE>$).*

We ignore the starting time of the event, considered less informative as the duration in our case.

Each SE is identified in the original dataset by a unique identifier called **SID**, and each of its events E has also an internal unique identifier called **EID** (an integer representing the **position** of E in SE). The length of SE, denoted by $|SE|$, is equal to the **number** of its events.

We assume that a total order exists on items denoted by $\prec$, which defines the order between events in the sequential event list. We note $E_1 \prec E_2$, if $\{E_1, E_2\} \in SE_i$ and $E_1.EID < E_2.EID$ (in other word, the event E_1 starts before the event E_2 in the sequential event list E_i).

Table 1. Example database of event lists and patterns detected

SID	Events (Label, Duration)	Event list (Label$_{eid}$)
1	E_1 (A, 3), E_2 (B, 6) E_3 (C, 7), E_4 (E, 9)	$EL_1 = A_1 \rightarrow B_2 \rightarrow C_3 \rightarrow E_4$
2	E_1 (A, 5), E_2 (B, 8) E_3 (C, 9)	$EL_2 = A_1 \rightarrow B_2 \rightarrow C_3$
3	E_1 (A, 3), E_2 (B, 4) E_3 (D, 1), E_4 (E, 3)	$EL_3 = A_1 \rightarrow B_2 \rightarrow D_3 \rightarrow E_4$
4	E_1 (A, 7), E_2 (B, 8) E_3 (D, 3)	$EL_4 = A_1 \rightarrow B_2 \rightarrow D_3$

Definition 3. *(subsequence). $SE' = \{E'_1, E'_2, ..., E'_n\}$ is called a subsequence of $SE = \{E_1, E_2, ..., E_m\}$, denoted by $SE' \subseteq SE$, if and only if there exists an injective function $f: SE'.events \rightarrow SE.events$, such as (1) $\forall E'_i \in SE' \rightarrow f(E'_i) \in SE$, (2) $E'_i.label = f(E'_i).label$, and (3) $\forall \{E'_i, E'_j\} \in SE'$, if $E'_i \prec E'_j \rightarrow f(E'_i) \prec f(E'_j)$. For example, $\{B, E\} \subset EL_3$, $EL_2 \subset EL_1$ and $EL_2 \not\subseteq EL_3$ can be deduced from Table 1.*

Definition 4. *(pattern). A **k-pattern** is a **frequent** k-subsequence, where k is its number of events (k = pattern.$|items|$).*

Table 2. Example of IDListExt built from Table 1 with minSupp = 2 (duration in minute)

A		
SID	EID	IDuration
1	1	3
2	1	5
3	1	3
4	1	7
INeighbors: { B, C, D, E }		

B		
SID	EID	IDuration
1	2	6
2	2	8
3	2	4
4	2	8
INeighbors: { C, D, E }		

C		
SID	EID	IDuration
1	3	7
2	3	9
INeighbors: {∅}		

D		
SID	EID	IDuration
3	3	1
4	3	3
INeighbors: {∅}		

E		
SID	EID	IDuration
1	4	9
3	4	3
INeighbors: {∅}		

Following [39], we represent the pattern in vertical format. A pattern, in the standard vertical format, has (1) **labels**: the sequence of its items' label, and (2) **IDList**: the list of positions (in the original dataset) where the pattern appears. Each position **Pos** in an $IDList$ is a couple of one SID and a list of EID. This vertical representation allows quick computing of the pattern's support which can be done by counting the number of all distinct SID in the $IDList$ of the pattern.

A valid *pattern* is constrained as follows:

1. frequency constraint: $support(pattern) \geq minSupp$, where $minSupp$ is a user-specified minimum support threshold
2. event order constraint: $\forall\{EID_i, EID_j\} \in Pos$, if $EID_i < EID_j$ then EID_i should appear before EID_j in the list of EID in Pos

In our case, the pattern's items should be coupled with the average duration spent on them. So, each item it in a pattern p should be coupled with a duration. This duration d_{it} is equal to the average of the time spent on it in all sequences where p appears. We choose to maintain d_{it} during the mining process. To achieve this goal, we augment the standard vertical representation by introducing **IDListExt**, an augmented vertical representation containing **IDurations**, **IFlags**, and **INeighbors** in addition.

Definition 5. *(IDListExt). An extension of IDList, which contains in addition: (1) IDurations: list of durations, (2) IFlags: list of flags, and (3) INeighbors: list of neighboring items.*

Definition 6. *(IDurations). List of durations of the last item of the pattern in each position where the pattern appears.*

To avoid data duplication, only the duration of the last item in the pattern is kept in the $IDurations$ of this pattern. To obtain the durations of the other items of the pattern, we can compute them from the parent pattern of the pattern in question. Refer to the Sect. 4 for more details.

Definition 7. *(Duration). The duration d_{it} of an item it in a pattern p can be defined as:*

$$d_{it} = \frac{\sum_{i=1}^{n}(duration\,of\,[it]\,in\,[SE_i])}{p.|SID|} \tag{1}$$

where (1) it $\in SE_i$ and (2) $p \subseteq SE_i$. For example, from Table 1, the pattern {A, B} can be represented with flags and durations by $P_{AB}=\{(A+,\ 4.5)\ (B,\ 6.5)\}$. With this representation, we can see that users spend on average 6.5 min on page B, when they consult page A before.

Definition 8. *(IFlags). List of flag for each item indicating if the duration of an item is longer or shorter than the previous one in the same pattern.*

To make the difference between two patterns P_1 and P_2 that have the same sequence of labels but a difference in the variation of durations, we choose to add a flag to each item in the pattern. This flag can have a value of "none", "+" or "−" according to the following rules: given a pattern $p = \{(I_i, D_i)(I_j, D_j)\}$, we set $flag(I_i)$ to:

1. "+": if $(D_j - D_i) \geq minGap$
2. "−": if $(D_j - D_i) < minGap$
3. "none": if I_i is the last item

where $minGap$ is a user-specified minimum gap threshold between consecutive items. For example, the patterns $P_1 = \{(A,2)(B,4)(C,6)\}$ and $P_2 = \{(A,9)(B,4)(C,6)\}$ can be represented with flags by $P_1 = \{(A+,2)(B+,4)(C,6)\}$ and $P_2 = \{(A-,9)(B+,4)(C,6)\}$ respectively for $minGap = 1$.

Two patterns having the same sequence of items' labels but a difference in the variation of duration should not be equal. This ensures that all patterns found have the same variation in duration. For example, from Table 1, $P_1=\{(B+,\ 6)(E,\ 9)\} \neq P_2=\{(B-,\ 4)\ (E,\ 3)\}$, where P_1 appears in sequence with SID=1 and P_2 appears in sequence with SID = 3. We notice that $P_i == P_j$ if and only if, the sequence of labels and flags in P_i is equal to the sequence of labels and flags in P_j respectively.

Definition 9. *(INeighbors). List of frequent items that appear after the last item of the pattern in the database. These neighbors are used in the candidate generation and avoid the generation of false candidates For example, in Table 1, for minSupp = 2, the neighbors of A are $\{< B, freq = 4 >, < C, freq = 2 >, < D, freq = 2 >, < E, freq = 2 >\}$, and those of B are $\{< C, freq = 2 >, < D, freq = 2 >\}$.*

A canonical representation of *pattern* can be obtained by merging all its frequent items (in the same order of its appearance) such as every item has a label, a flag and an average duration. Two patterns $k-P_1$ and $k-P_2$ are equal if they have the same sequence of label, and flags such as: (1) $P_1.|items| = P_2.|items| = k$, and for any $1 \leq i \leq k$, we have (2) $P_1.Item_i.label = P_2.Item_i.label$ and (3) $P_1.Item_i.flag = P_2.Item_i.flag$.

4   Sequential Event Pattern Mining

Let $D = \{SE_1, ..., SE_n\}$ be an input database of SE, $minSupp$ a user-specified minimum support threshold. The goal is to find all sequential patterns appearing at least in $minSupp$ sequences in D, and including the average duration of their items. In this section, we describe our proposed algorithms to (1) build the IDListExt of all frequent events and (2) extract the sequential patterns. These algorithms are inspired by the algorithm proposed in [21] called **Sequential Event Pattern Mining (SEPM)**.

Three extended algorithms are proposed for the mining: **SEPM−** (Algorithm 5) for mining sequential patterns without duration, **SEPM+** (Algorithm 1) for mining sequential patterns with duration, and **SEPM++** (Algorithm 2) for mining sequential patterns with duration in parallel (multithread). All these algorithms use the Depth-First Search to mine the patterns. Refer to the Appendix A for more details.

4.1   Build the List of IDListExt

Algorithm 3 is proposed to build the IDListExt of each frequent event in the database. It starts by scanning the database to obtain all single frequent events $freqEvents$ (infrequent events can be removed from the database to accelerate the second scan). Then, a second scan is performed to build the IDListExt of each frequent event. The last operation will remove all infrequent neighbors from all INeighbor.

4.2   SEPM Without Duration

Algorithm 5, called **SEPM−**, is proposed to extract patterns without duration. The resulting patterns are used to improve the utility of duration (refer to Sect.5 for more details). SEPM− starts by generating the list of candidates. Then, it computes the frequency of each candidate, and if it is greater then or equal to the $minSupp$ the candidate becomes a pattern. For all (k)-patterns generated, the SEPM− repeats the process to explore the $(k+1)$-patterns. This recursive process is repeated until all patterns are found. An intersection $\cap$ between two IDListExt is performed to retrieve the positions of the candidate generated. At the beginning of the intersection step, all the common SID (between the two IDListExt) are computed. Then for each of them, we search in $neighbor$.IDListExt, the minimum EID which is greater than the max EID in the $pattern$.IDListExt. If this minimum exists, it means that the candidate exists in the sequence with id equal to SID.

4.3   SEPM with Duration

Algorithm 5, called **SEPM+**, is proposed to extract patterns with duration. It has the same steps as the previous Algorithm 5 except the intersection $\cap$

between the IDListExt. Here, the intersection can generate patterns (from 0 to 2) according to the variation in the duration at each common SID. Items' flags of the patterns are used to encode the duration variation. When a pattern joins its neighbor, the variation between the duration of this neighbor and the duration of the last item in the pattern in each common SID will determine the flag of the last item in the pattern (before the addition of the neighbor to the pattern). For example, joining $\{A+B+C\}$ with $\{D\}$ can produce $\{\emptyset\},\{A+B+C+D\}$, and/or $\{A+B+C-D\}$.

Algorithm 1: SEPM+

Input: Pattern: *pattern*
Input: User-specified threshold: *minSupp*
Input: Set of IDListExt: ν
1 save(*pattern*)
2 **forall the** *neighbor* $\in$ *pattern.INeighbors* **do**
3 $\quad$ $IDListExt \leftarrow pattern.IDListExt \cap \nu$.find(*neighbor*).$IDListExt$
4 $\quad$ **if** $IDListExt.|SID| \geq minSupp$ **then**
5 $\quad\quad$ $newPatterns \leftarrow$
$\quad\quad$ createPatternsWithDuration(*pattern*, *neighbor*, $IDListExt$)
6 $\quad\quad$ **forall the** *newPattern* $\in$ *newPatterns* **do**
7 $\quad\quad\quad$ $|$ SEPM+(*newPattern*) // `recursive call`
8 $\quad\quad$ **end**
9 $\quad$ **end**
10 **end**

An example is shown in Fig. 1. It shows how patterns and their ID-Lists are generated from a dataset.

4.4 SEPM with Parallel Mining Process

Algorithm 2, called **SEPM++**, is proposed to extract patterns with duration in parallel to accelerate the mining process. The number of threads used can be adjusted according to the number of available processing units. The output is the same as for SEPM+. At the beginning a thread pool is created at line 1, then all 1-patterns are extracted at line 2. After that, a task is created for each 1-pattern and added to the task list of the thread pool at line 3–6. When a task is added, it will be directly executed by the first available thread. A thread generates all possible patterns from the 1-pattern of a given task (like in SEPM+). Finally, the execution of the SEPM++ will be paused as long as there are running threads and unprocessed tasks at line 7.

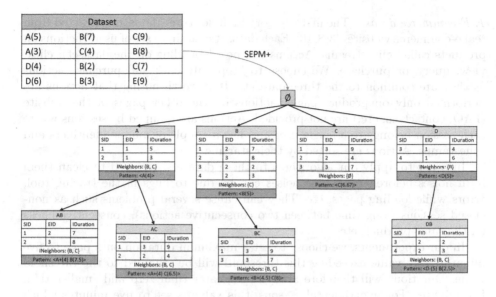

Fig. 1. Tree showing the pattern tree extracted from the dataset in this figure.

Algorithm 2: SEPM++

Input: User-specified threshold: $minSupp$
Input: Number of threads: $nbThread$
Input: Set of IDListExt: ν // **of all frequent events**
1 $t\_pool \leftarrow$ create a pool of $nbThread$ threads
2 $1\_patterns \leftarrow$ build all 1-$pattern$ from ν
3 **forall the** $pattern \in 1\_patterns$ **do**
4 | $task \leftarrow$ create a task from $pattern$
5 | $pool$.add_task($task$)
6 **end**
7 wait all thread $pool$
8 shutdonw all threads in $pool$

---

5   Experimental Results

Our proposed algorithms was applied (1) on several real **e-commerce** datasets to demonstrate the relevance of the mined patterns and (2) on **synthetic** datasets to prove their efficiency and scalability. All experiments were performed on a computer with Core-i5 with 5 physical cores, running on Windows 10 and 16 GB of RAM.

5.1   Datasets

In this section, we describe the two categories of datasets used in our experiments, and how they are constructed.

1. E-commerce dataset The first category includes three datasets extracted from real e-commerce websites [3, 8, 10]. Each dataset contains a set of users' actions on products called clickstreams. Actions can be (depending on the dataset): click, view, query, or purchase. We choose to keep only click and purchase actions as they are common to the three datasets. It is worth noting that actions are performed only on product pages, actions on the other pages of the website (FAQ, contact-us, etc.) are not provided. Actions are grouped by sessions where each session belongs to a single user and has a couple of unique identifiers and timestamps. Actions are sorted by their starting time.

In order to apply our algorithms to these datasets, we need to clean them from noises beforehand. These noises can be due to bugs in the tracing tool, errors while loading pages, etc. They can cause several problems such as non-closed sessions (long time between two consecutive actions), consecutive clicks on the same product, etc.

In our experiments, we choose to set a maximum duration on a page to five minutes. Any value exceeding this threshold will be trimmed to the threshold value. Durations will therefore always be greater than zero and smaller than five minutes. For experimental reasons, this value is set to five minutes which equal to the third quartile (75th percentile) of the values of the durations in the datasets. In an ideal case, detect outliers algorithms should be used to eliminate the outliers without having to set a maximum duration value. We have not addressed this point as it was not our priority in this paper.

After cleaning, clickstreams should be converted into sequential event data. Every session is converted into an event-list and every click in a session is converted into an event attached to its corresponding event-list. The duration of an event E_i is defined as the time between the beginning of two consecutive events (E_i and E_{i+1}) in the same session. All the clicks durations in a session can be obtained except for the duration of the last one. This duration is replaced by the average duration of all clicks in the current session.

Table 3 contains some statistics about the three clickstream datasets, and Table 4 contains the number and length of patterns founded by SEPM+ in each dataset.

Table 3. Summary of properties of the e-commerce datasets. E: Events. SE: Sequential events

Dataset	|SE|	|E|	|Distinct E|	Mean E in SE	Session with purchase
CIKM	238k	2,121k	75k	3.11	12k
YooChoose	670k	3,87k	45k	3.77	410k
AliBaba	1,119k	13,966k	4,085k	6.46	630k

Sample file (1,119k line)

Table 4. Statistics of patterns found using SEPM+ in each dataset. The number of patterns found may be different from those found by SEPM as the representation of patterns is different. Datasets contain sessions with and without purchase. The number of patterns found with SEPM+ is not equal to that found by SEPM as the representation of patterns is not the same

Dataset	minSupp	Nb patterns	Max length
CIKM	0.01%	23,221	4
YooChoose	0.01%	26,807	8
AliBaba	0.01%	8,461	6

2. Synthetic dataset The synthetic dataset was generated using *IBM data quest generator*. They are used to study the effect of varying properties of the input database on the execution time. The generation of datasets can be controlled through several parameters:

- **ncrust** (number of customers in the database) to vary the number of event lists (i.e. **D**)
- **slen** (average of transactions per customer) to vary the number of events per list (i.e. **L**)
- **nitems** (the number of different items available) to vary the number of types (labels) of events (i.e. **T**)

Three synthetic datasets are generated by varying each parameter.

- For D, we set *minSupp* to 10% and we vary D between 250k and 400k (dataset **SYN1-4**).
- For L, *minSupp* is set to 10% and L is varied between 8 and 15 (dataset **SYN5-8**).
- For T, *minSupp* is set to 5% and T is varied between 3 and 10 (dataset **SYN9-12**). Table 5 shows some statistics about the three synthetic datasets generated.
- Lastly, dataset **SYN13** was generated to evaluate the performance of the algorithms with respect to the minSupp.

5.2   Categorization

Some of the 4-patterns found are presented in Table 7. We can observe that these patterns are more instructive than those containing only labels since they also contain the duration. To benefit from these resulting patterns and to analyze them, we propose to categorize them. We notice that durations are not always equivalent, they are different from an item to another in the same pattern (see Table 7). We also notice a difference in the variation of the duration in the same pattern. Based on these two observations, we decide to categorize the patterns

Table 5. A summary of the synthetic datasets. E: Events. SE: Sequential events

Dataset	|SE|	|E|	|Distinct E|	Mean E in SE
SYN1	250k	7,630k	3,240	30
SYN2	300k	9,160k	3,240	30
SYN3	350k	10,680k	3,240	30
SYN4	400k	12,215k	3,240	30
SYN5	300k	7,080k	3,240	25
SYN6	300k	9,160k	3,240	30
SYN7	300k	12,200k	3,240	40
SYN8	300k	14,220k	3,240	50
SYN9	300k	9,160k	2,250	30
SYN10	300k	9,160k	4,350	30
SYN11	300k	9,160k	5,490	30
SYN12	300k	9,160k	6,140	30
SYN13	300k	9,160k	3,240	30

according to two properties: the standard deviation of durations called **sd**, and the number of variations of durations called **variation**, such as:

$$sd_p = \sqrt{\frac{\sum_{i=1}^{l_p}(d_i - v)^2}{l_p}} \tag{2}$$

where l_p is the length of pattern **p**, d_i is the duration of its ith item, and v is the mean of all durations in p,

$$variation_p = \sum_{i=1}^{l_p-1}(\mathbf{diff}(flag_i, flag_{i+1})) \tag{3}$$

where $flag_i$ is the flag of the ith item in p and $diff(x, y)$ is equal to 0 if x is equal to y and 1 otherwise.

To detect abnormal behaviors, we use two user-specified thresholds $maxVariation$ and $maxSD$ to filter them. In our tests, we set $maxVariation$ to 10, $maxSD$ to 2, and we keep only patterns with minimum of 4 items. Algorithm 8 describes the main categorization steps and Table 6 shows some statistics on the different categories detected.

Based on these categories, mining techniques could be applied[1] to better understand the customers' behaviors. E.g., customers can be categorized based on frequent pattern categories in their sessions, and then each customer's profile can be treated separately. Various customers' demographic data could be combined with the patterns categories to recommend more personalized websites.

---

[1] These mining techniques are not addressed in our work because of the lack of public data containing this information.

Table 6. Statistics on the resulting categories. var = i means variation is equal to i. A sample file is used for AliBaba dataset

Dataset	var = 0	var = 1	var = 2	var = 3	...	Abnormal
CIKM	12,539	6,170	4,200	115	...	23
YooChoose	16,734	8,127	551	448	...	119
AliBaba	5,091	1,304	660	525	...	145

Table 7. Example of 4-patterns found

Label of product (duration in minutes)			
9344+(3.17) $\rightarrow$	9344−(4.5) $\rightarrow$	3599+(1.99) $\rightarrow$	14458(5)
3785+(0.69) $\rightarrow$	32320+(2.8) $\rightarrow$	8848+(4) $\rightarrow$	35358(5)
4352300−(2.04) $\rightarrow$	2790543+(0.17) $\rightarrow$	5051027+(0.47) $\rightarrow$	2859111(1.58)
4960783+(0.17) $\rightarrow$	2081505+(0.15) $\rightarrow$	2020265+(0.21) $\rightarrow$	4614885(0.22)

5.3 Discriminatory Patterns

In this section, we evaluate the effectiveness of patterns with duration mined with SEPM+. The goal in this experiment is to prove that by using the duration, we can make the patterns more discriminating than those without duration. In this experiment, we use SEPM+ and SEPM−[2] to detect patterns with duration and patterns without duration respectively.

Preparing the Patterns. We start by splitting the clickstream into two groups: the first one containing users' sessions with at least one purchase and the second one those who do not contain any purchase. Then, by using SEPM+ and SEPM−, we mine the patterns in each group. With each algorithm, we obtain patterns that lead to a purchase called **P+** and patterns that do not lead to a purchase called **P−**. For the correct use of these patterns, the non-discriminatory patterns called P_{non_disc} should be removed. P_{non_disc} are the patterns that exist in both P+ and P−. So, the goal is to prove that by using SEPM+, the non-discriminatory patterns detected with SEPM− become discriminatory by adding the duration. For example, given a pattern $P_{non\_disc1} = \{A, B\}$ which exists in both P+ and P−, it is difficult to assume whether P_{non_disc1} represents purchase behaviors or not, since patterns are represented by only labels (results of SEPM−). Contrariwise, with our proposed algorithm SEPM+, the duration added to the patterns will be used to separate the patterns that have the same sequence of the label but not the same sequence of flags (see Fig. 2).

---

[2] Any other classic sequential pattern mining algorithm (like SPADE, SPAM, etc.) that can be used since they all have the same output/patterns.

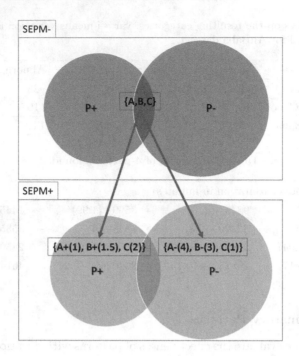

Fig. 2. $\{A, B\}$ is a non-discriminatory pattern in SEPM−, but it is not the case in SEPM+ as the patterns are represented differently

Effectiveness. To measure the added duration in discrimination power, we start by (1) removing all patterns that have only one item as they are not very relevant in the analysis. Then (2) we detect patterns P_{non_disc} in SEPM− that exist in both P+ and P− (see Fig. 2). Finally, (3) we will verify if P_{non_disc} (found previously) exist in P+ or P− of SEPM+. By doing that, we check if the non-discriminatory patterns rejected by SEPM− become discriminatory with SEPM+.

Results. Table 8 shows the results of this experiment. As we can see, all non-discriminatory patterns rejected by an algorithm without duration (during the analysis phase) become discriminatory with SEPM+. It proves that we successfully transform non-discriminatory patterns into discriminatory ones by adding the duration into the pattern representation.

Table 8. Effectiveness of the added duration, where *effectiveness* is the percentage of non-discriminatory patterns rejected by SEPM− and kept by SEPM+

Algorithm used	Parameters & results	CIKM	YooChoose	AliBaba
SEPM+/SEPM−	MinSupp	4 SE	4 SE	4 SE
SEPM+/SEPM−	MinGap	1.5 min	1.5 min	1.5 min
SEPM+	NbPatterns(with purchase)	5k	93k	14k
SEPM+	NbPatterns(without purchase)	72k	209k	13k
SEPM−	NbPatterns(with purchase)	5k	175k	22k
SEPM−	NbPatterns(without purchase)	87k	282k	18k
SEPM+	P_{non_disc} (rejected)	8%	2%	1%
SEPM−	P_{non_disc}	88%	31%	9%
	Effectiveness	**98%**	**97%**	**99%**

Sample file (150k line)

5.4 Performances

In this section, we apply seven algorithms (SEPM, SPEM[−,+,++], PrefixS-pan, SPAM, and SPADE) on synthetic datasets. The source codes of algorithms (PrefixSpan, SPAM, and SPADE) are provided by [14]. Figures 3, 4, and 5 shows the time response of these algorithms according to some dimensions of the dataset. In all test except the last one, 5 threads are used to run SEPM++. Figure 6 shows the response time of these algorithms when *minSupp* varies and Fig. 7 shows the response time of SEPM++ according to the number of threads

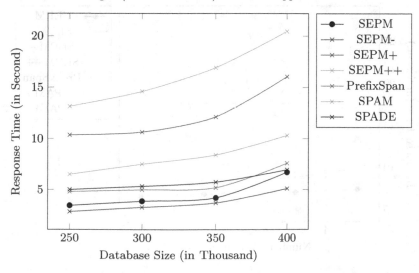

Fig. 3. The effect of varying the length of the database

used. As we can see, in all cases, our proposed algorithms behave in the same way as the other algorithms when properties of the database change. Results show that our algorithms behave correctly according to the different types of databases.

Number of events per list (dataset SYN5-8) with minSupp=10%

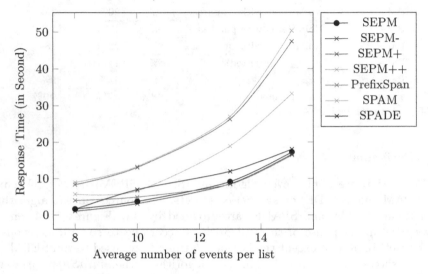

Fig. 4. The effect of varying the number of events per list

Label of events (dataset SYN9-12) with minSupp=5%

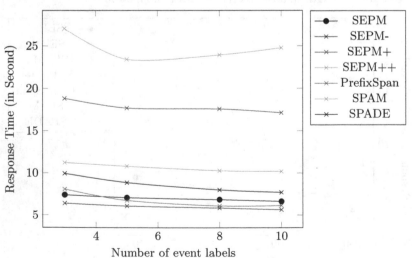

Fig. 5. The effect of varying the number of event labels

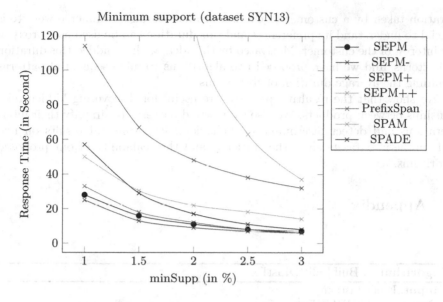

Fig. 6. Effect of varying the minSupp

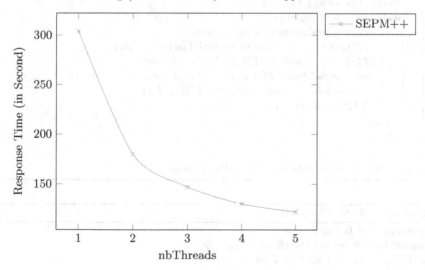

Fig. 7. Effect of varying the number of threads

6   Conclusion

E-sellers attempt to understand their customers' behaviors and discover how they navigate through their e-commerce website, why they select a particular product, or even why they abandon their purchasing process. We believe that the duration taken by a customer to check a product on an e-commerce website is crucial to understand its preferences and this duration can be used as a proxy of the interest of the customer. Motivated by this idea, we have added this duration to patterns, and we have proposed two algorithms to mine sequential patterns including the average duration of their items.

We show that the resulting patterns are useful for discovering hidden relationships between products. We also categorized patterns to simplify their interpretations and detect discriminatory behaviors. Experimental results on real and synthetic datasets show the efficiency and the scalability of our proposed algorithms.

A   Appendix

Algorithm 3: Build all IDListExt

Input: Event List: db
Output: Set of IDListExt: ν
1 $freqEvents \leftarrow$ all labels of frequent items
2 **forall the** $el \in db$ **do**
3      $antecedents \leftarrow \emptyset$, $EID \leftarrow 1$
4   **forall the** $e \in el.events$ **do**
5   **if** $freqEvents$ *contains e.label* **then**
6              $IDListExt \leftarrow \nu$.findOrCreateIDListExt($e.label$)
7              $IDListExt$.add($el.SID, EID, e.duration$)
8   add $e.label$ to all $IDListExt.INeighbor$ in $antecedents$
9              $antecedents \leftarrow antecedents \cup IDListExt$
10             $EID \leftarrow EID + 1$
11   **end**
12   **end**
13 **end**
14 remove infrequent neighbors from all INeighbors

Algorithm 4: SEPM Main

Input: Set of IDListExt: ν
Input: User-specified threshold: $minSupp$
1 $patterns \leftarrow$ build 1-$pattern$ from ν
2 **forall the** $p \in patterns$ **do**
3   <SEPM-/SEPM+>($p, minSupp, \nu$)
4 **end**

Algorithm 5: SEPM-

Input: Pattern: *pattern*
Input: User-specified threshold: *minSupp*
Input: Set of IDListExt: ν

1 save(*pattern*)
2 **forall the** *neighbor* $\in$ *pattern.INeighbors* **do**
3 $\quad$ *IDListExt* $\leftarrow$ *pattern.IDListExt* $\cap$ ν.find(*neighbor*).*IDListExt*
4 $\quad$ **if** *IDListExt.*|*SID*| $\geq$ *minSupp* **then**
5 $\quad\quad$ *newPattern* $\leftarrow$ createPattern(*pattern, neighbor, IDListExt*)
6 $\quad\quad$ SEPM-(*newPattern*) // **recursive call**
7 $\quad$ **end**
8 **end**

Algorithm 6: SEPM+

Input: Pattern: *pattern*
Input: User-specified threshold: *minSupp*
Input: Set of IDListExt: ν

1 save(*pattern*)
2 **forall the** *neighbor* $\in$ *pattern.INeighbors* **do**
3 $\quad$ *IDListExt* $\leftarrow$ *pattern.IDListExt* $\cap$ ν.find(*neighbor*).*IDListExt*
4 $\quad$ **if** *IDListExt.*|*SID*| $\geq$ *minSupp* **then**
5 $\quad\quad$ *newPatterns* $\leftarrow$
$\quad\quad$ createPatternsWithDuration(*pattern, neighbor, IDListExt*)
6 $\quad\quad$ **forall the** *newPattern* $\in$ *newPatterns* **do**
7 $\quad\quad\quad$ SEPM+(*newPattern*) // **recursive call**
8 $\quad\quad$ **end**
9 $\quad$ **end**
10 **end**

Algorithm 7: SEPM++

Input: User-specified threshold: *minSupp*
Input: Number of threads: *nbThread*
Input: Set of IDListExt: ν // **of all frequent events**

1 *t_pool* $\leftarrow$ create a pool of *nbThread* threads
2 1_*patterns* $\leftarrow$ build all 1-*pattern* from ν
3 **forall the** *pattern* $\in$ 1_*patterns* **do**
4 $\quad$ *task* $\leftarrow$ create a task from *pattern*
5 $\quad$ *pool*.add_task(*task*)
6 **end**
7 wait all thread *pool*
8 shutdonw all threads in *pool*

Algorithm 8: Categorization

   Input: Patterns: *patterns*
   Input: User-specified threshold: $maxSD$
   Input: User-specified threshold: $maxVariation$
1 $categories \leftarrow \emptyset$
2 **forall the** *pattern* $\in$ *patterns* **do**
3     $pVariation \leftarrow pattern.variation()$
4     $pSD \leftarrow pattern.sd()$
5     $index \leftarrow -1$
6   **if** $pVariation < maxVariation$ **and** $pSD < maxSD$ **then**
7        $index \leftarrow$ find_category($npVariation$)
8   **else**
9        $index \leftarrow 0$ // First category is reserved to the abnormal patterns
10   **end**
11   add *pattern* to *categories*[*index*]
12 **end**

References

1. Agrawal, R., Srikant, R.: Mining sequential patterns. In: ICDE, p. 3. IEEE (1995)
2. Alborzi, M., Khanbabaei, M.: Using data mining and neural networks techniques to propose a new hybrid customer behaviour analysis and credit scoring model in banking services based on a developed RFM analysis method. Int. J. Bus. Inf. Syst. **23**(1), 1–22 (2016)
3. Alibaba: (dataset) user behavior data from Taobao for recommendation (2018). https://tianchi.aliyun.com/dataset/dataDetail?dataId=649
4. Allen, J.F.: Maintaining knowledge about temporal intervals. Commun. ACM **26**, 832–843 (1983)
5. Alzahrani, M.Y., Mazarbhuiya, F.A.: Discovering sequential patterns from medical datasets. In: 2016 International Conference on Computational Science and Computational Intelligence (CSCI), pp. 70–74. IEEE (2016)
6. Ansari, A., Riasi, A.: Customer clustering using a combination of fuzzy c-means and genetic algorithms. Int. J. Bus. Manage. **11**(7), 59 (2016)
7. Ayres, J., Flannick, J., Gehrke, J., Yiu, T.: Sequential pattern mining using a bitmap representation. In: Proceedings of the Eighth ACM SIGKDD International Conference on Knowledge Discovery and Data Mining, pp. 429–435. ACM (2002)
8. Ben-Shimon, D., Tsikinovsky, A., Friedmann, M., Shapira, B., Rokach, L., Hoerle, J.: RecSys challenge 2015 and the YOOCHOOSE dataset. In Proceedings of the 9th ACM Conference on Recommender Systems, pp. 357–358. ACM (2015)
9. Chen, K.-Y., Jaysawal, B.P., Huang, J.-W., Wu, Y.-B.: Mining frequent time interval-based event with duration patterns from temporal database. In: 2014 International Conference on Data Science and Advanced Analytics (DSAA), pp. 548–554. IEEE (2014)
10. CIKM: (dataset) CIKM CUP 2016 track 2: Personalized e-commerce search challenge (2016). https://competitions.codalab.org/competitions/11161

11. Dursun, A., Caber, M.: Using data mining techniques for profiling profitable hotel customers: an application of RFM analysis. Tour. Manage. Perspect. **18**, 153–160 (2016)
12. Fournier-Viger, P., Gomariz, A., Campos, M., Thomas, R.: Fast vertical mining of sequential patterns using co-occurrence information. In: Tseng, V.S., Ho, T.B., Zhou, Z.-H., Chen, A.L.P., Kao, H.-Y. (eds.) PAKDD 2014. LNCS (LNAI), vol. 8443, pp. 40–52. Springer, Cham (2014). https://doi.org/10.1007/978-3-319-06608-0_4
13. Fournier-Viger, P., Wu, C.-W., Gomariz, A., Tseng, V.S.: VMSP: efficient vertical mining of maximal sequential patterns. In: Sokolova, M., van Beek, P. (eds.) AI 2014. LNCS (LNAI), vol. 8436, pp. 83–94. Springer, Cham (2014). https://doi.org/10.1007/978-3-319-06483-3_8
14. Fournier-Viger, P., et al.: The SPMF open-source data mining library version 2. In: Berendt, B., et al. (eds.) ECML PKDD 2016. LNCS (LNAI), vol. 9853, pp. 36–40. Springer, Cham (2016). https://doi.org/10.1007/978-3-319-46131-1_8
15. Fournier-Viger, P., Lin, C.-W., Kiran, R.U., Koh, Y.S., Thomas, R.: A survey of sequential pattern mining. Data Sci. Pattern Recogn. **1**(1), 54–77 (2017)
16. García-Hernández, R.A., Martínez-Trinidad, J.F., Carrasco-Ochoa, J.A.: A new algorithm for fast discovery of maximal sequential patterns in a document collection. In: Gelbukh, A. (ed.) CICLing 2006. LNCS, vol. 3878, pp. 514–523. Springer, Heidelberg (2006). https://doi.org/10.1007/11671299_53
17. Gomariz, A., Campos, M., Marin, R., Goethals, B.: ClaSP: an efficient algorithm for mining frequent closed sequences. In: Pei, J., Tseng, V.S., Cao, L., Motoda, H., Xu, G. (eds.) PAKDD 2013. LNCS (LNAI), vol. 7818, pp. 50–61. Springer, Heidelberg (2013). https://doi.org/10.1007/978-3-642-37453-1_5
18. Huaulmé, A., Voros, S., Riffaud, L., Forestier, G., Moreau-Gaudry, A., Jannin, P.: Distinguishing surgical behavior by sequential pattern discovery. J. Biomed. Inf. **67**, 34–41 (2017)
19. Jagan, S., Rajagopalan, S.P.: A survey on web personalization of web usage mining. Int. Res. J. Eng. Technol. **2**(1), 6–12 (2015)
20. Jia, R., Li, R., Yu, M., Wang, S.: E-commerce purchase prediction approach by user behavior data. In: 2017 International Conference on Computer, Information and Telecommunication Systems (CITS), pp. 1–5. IEEE (2017)
21. Kanaan, M., Kheddouci, H.: Mining patterns with durations from e-commerce dataset. In: Aiello, L.M., Cherifi, C., Cherifi, H., Lambiotte, R., Lió, P., Rocha, L.M. (eds.) COMPLEX NETWORKS 2018. SCI, vol. 812, pp. 603–615. Springer, Cham (2019). https://doi.org/10.1007/978-3-030-05411-3_49
22. Kontostathis, A., Galitsky, L.M., Pottenger, W.M., Roy, S., Phelps, D.J.: A survey of emerging trend detection in textual data mining. In: Berry, M.W. (eds.) Survey of Text Mining, pp. 185–224. Springer, New York (2004). https://doi.org/10.1007/978-1-4757-4305-0_9
23. Liao, V.C.-C., Chen, M.-S.: DFSP: a Depth-First SPelling algorithm for sequential pattern mining of biological sequences. Knowl. Inf. Syst. **38**(3), 623–639 (2013). https://doi.org/10.1007/s10115-012-0602-x
24. Lin, N.P., Hao, W.-H., Chen, H.-J., Chueh, H.-E., Chang, C.-I., et al.: Fast mining of closed sequential patterns. WSEAS Trans. Comput. **7**(3), 1–7 (2008)
25. Mabroukeh, N.R., Ezeife, C.I.: A taxonomy of sequential pattern mining algorithms. ACM Comput. Surv. (CSUR) **43**(1), 3 (2010)
26. Martínez, A., Schmuck, C., Pereverzyev Jr., S., Pirker, C., Haltmeier, M.: A machine learning framework for customer purchase prediction in the non-contractual setting. Eur. J. Oper. Res. **281**(3), 588–596 (2018)

27. Mobasher, B., Dai, H., Luo, T., Nakagawa, M.: Using sequential and non-sequential patterns in predictive web usage mining tasks. In: 2002 IEEE International Conference on Data Mining. Proceedings, pp. 669–672. IEEE (2002)
28. Najafabadi, M.K., Mahrin, M.N.R., Chuprat, S., Sarkan, H.M.: Improving the accuracy of collaborative filtering recommendations using clustering and association rules mining on implicit data. Comput. Hum. Behav. **67**, 113–128 (2017)
29. Neysiani, B.S., Soltani, N., Mofidi, R., Nadimi-Shahraki, M.H.: Improve performance of association rule-based collaborative filtering recommendation systems using genetic algorithm. Int. J. Inf. Technol. Comput. Sci. **2**, 48–55 (2019)
30. Patel, D., Hsu, W., Lee, M.L.: Mining relationships among interval-based events for classification. In: Proceedings of the 2008 ACM SIGMOD International Conference on Management of Data, pp. 393–404. ACM (2008)
31. Patil, S.S., Khandagale, H.P.: Survey paper on enhancing web navigation usability using web usage mining techniques. Int. J. Mod. Trends Eng. Res. (IJMTER) **3**(02), 594–599 (2016)
32. Pei, J., et al.: Mining sequential patterns by pattern-growth: the PrefixSpan approach. IEEE Trans. Knowl. Data Eng. **16**(11), 1424–1440 (2004)
33. Srikant, R., Agrawal, R.: Mining sequential patterns: generalizations and performance improvements. In: Apers, P., Bouzeghoub, M., Gardarin, G. (eds.) EDBT 1996. LNCS, vol. 1057, pp. 1–17. Springer, Heidelberg (1996). https://doi.org/10.1007/BFb0014140
34. Srivastava, J., Cooley, R., Deshpande, M., Tan, P.-N.: Web usage mining: discovery and applications of usage patterns from web data. ACM SIGKDD Explor. Newslett. **1**(2), 12–23 (2000)
35. Tóth, K., Kósa, I., Vathy-Fogarassy, Á.: Frequent treatment sequence mining from medical databases. Stud. Health Technol. Inf. **236**, 211–218 (2017)
36. Wang, K., Xu, Y., Yu, J.X.: Scalable sequential pattern mining for biological sequences. In Proceedings of the Thirteenth ACM International Conference on Information and Knowledge Management, pp. 178–187. ACM (2004)
37. Wu, Y., Ester, M.: FLAME: a probabilistic model combining aspect based opinion mining and collaborative filtering. In: Proceedings of the Eighth ACM International Conference on Web Search and Data Mining, pp. 199–208. ACM (2015)
38. Yates, A., Kolcz, A., Goharian, N., Frieder, O.: Effects of sampling on Twitter trend detection. In: Proceedings of the Tenth International Conference on Language Resources and Evaluation, LREC 2016, pp. 2998–3005 (2016)
39. Zaki, M.J.: SPADE: an efficient algorithm for mining frequent sequences. Machine learning **42**(1–2), 31–60 (2001). https://doi.org/10.1023/A:1007652502315
40. Zaman, T.S., Islam, N., Ahmed, C.F., Jeong, B.S.: iWAP: a single pass approach for web access sequential pattern mining. GSTF J. Comput. (JoC) **2**(1), 1–6 (2018)
41. Zeng, M., Cao, H., Chen, M., Li, Y.: User behaviour modeling, recommendations, and purchase prediction during shopping festivals. Electron. Markets **29**(2), 263–274 (2018). https://doi.org/10.1007/s12525-018-0311-8
42. Zignani, M., Quadri, C., Del Vicario, M., Gaito, S., Rossi, G.P.: Temporal communication motifs in mobile cohesive groups. In: Cherifi, C., Cherifi, H., Karsai, M., Musolesi, M. (eds.) COMPLEX NETWORKS 2017 2017. SCI, vol. 689, pp. 490–501. Springer, Cham (2018). https://doi.org/10.1007/978-3-319-72150-7_40

Scalable Schema Discovery for RDF Data

Redouane Bouhamoum$^{(\boxtimes)}$, Zoubida Kedad, and Stéphane Lopes

DAVID lab., University of Versailles Saint-Quentin-en-Yvelines, Versailles, France
{redouane.bouhamoum,zoubida.kedad,stephane.lopes}@uvsq.fr

Abstract. The semantic web provides access to an increasing number of linked datasets expressed in RDF. One feature of these datasets is that they are not constrained by a schema. Such schema could be very useful as it helps users understand the structure of the entities and can ease the exploitation of the dataset. Several works have proposed clustering-based schema discovery approaches which provide good quality schema, but their ability to process very large RDF datasets is still a challenge. In this work, we address the problem of automatic schema discovery, focusing on scalability issues. We introduce an approach, relying on a scalable density-based clustering algorithm, which provides the classes composing the schema of a large dataset. We propose a novel distribution method which splits the initial dataset into subsets, and we provide a scalable design of our algorithm to process these subsets efficiently in parallel. We present a thorough experimental evaluation showing the effectiveness of our proposal.

Keywords: Schema discovery · RDF Data · Clustering · Big Data

1 Introduction

The web of data represents a huge information space consisting of an increasing number of interlinked datasets described using languages proposed by the W3C such as RDF, RDFS and OWL. The Resource Description Framework (RDF)[1] is a standard model for data creation and publication on the web, while RDF Schema (RDFS)[2] was introduced to define a vocabulary which can be used to describe an RDF dataset. The Ontology Web Language (OWL)[3] is designed to represent rich and complex knowledge related to an RDF dataset. OWL documents are known as ontologies.

One important feature of such datasets is that they contain both the data and the schema describing the data. A good practice for the dataset publisher is to provide schema related declarations, such as the VoID's predicates[4], which capture various metadata describing a source. These declarations help the users

---

[1] RDF: https://www.w3.org/RDF/.
[2] RDFS: https://www.w3.org/TR/rdf-schema/.
[3] OWL: https://www.w3.org/OWL/.
[4] VoID: The Vocabulary of Interlinked Datasets.

© Springer-Verlag GmbH Germany, part of Springer Nature 2020
A. Hameurlain and A M. Tjoa (Eds.): TLDKS XLVI, LNCS 12410, pp. 91–120, 2020.
https://doi.org/10.1007/978-3-662-62386-2_4

understand the nature of the entities within an RDF dataset. However, these schema-related declarations are not mandatory, and they are not always provided. As a consequence, the schema may be incomplete or missing. Furthermore, even if the schema is provided, data are not constrained by this schema: resources of the same type may be described by property sets which are different from those specified in the schema.

The lack of schema offers a high flexibility while creating interlinked datasets, but can also limit their use. Indeed, it is not easy to query or explore a dataset without any knowledge about its resources, classes or properties. The exploitation of an RDF dataset would be straightforward with a schema describing the data. In the context of web data, a schema is viewed as a guide easing the exploitation of the RDF dataset, and not as a structural constraint over the data.

Several works have focused on schema discovery for RDF datasets. Some of these works rely on clustering algorithms to automatically extract the underlying schema of an RDF dataset [9,17,18]. These approaches explore instance-level data in order to infer a schema providing the classes and properties which describe the instances in the dataset. While these schema discovery approaches succeed in providing a good quality schema, their scalability is still an open issue as they rely on costly clustering algorithms. The use of such algorithms for discovering the underlying schema of massive datasets remains challenging due to their complexity.

In our work, we have addressed this scalability issue. Our goal is to propose a schema discovery approach suitable for very large datasets. To this end, we introduce in this paper a scalable density-based clustering algorithm specifically designed for schema discovery in large RDF datasets. Our approach parallelizes the clustering process and ensures that the result is the same as the one provided by a sequential algorithm. The main contributions presented in this paper are the followings:

- A novel distribution method dividing the initial dataset into subsets which can be processed efficiently in parallel, as well as an optimization of this method which limits the size of the subsets, thus limiting the number of comparisons among entities during the clustering.
- A parallel clustering algorithm suitable for a distributed environment which limits the costly information exchange operations between the calculating nodes.
- A scalable implementation of our algorithm based on the distributed processing framework Apache Spark[29], with the source code available online[5].
- A thorough experimental evaluation illustrating both the quality of the discovered classes and the performances of our approach.

This paper is organized as follows. The motivation behind our proposal is presented in Sect. 2. A global overview of our approach is provided in Sect. 3. Data distribution is detailed in Sect. 4, and neighbor identification is described in Sect. 5. Section 6 presents the local clustering process and Sect. 7 describes the

---

[5] https://github.com/BOUHAMOUM/SC-DBSCAN.

merging stage which produces the final clustering result. Experimental results are presented in Sect. 8. Section 9 discusses the existing approaches for schema discovery. Finally, Sect. 10 concludes the paper and presents our future works.

2 Motivation

In the web of data, datasets are created using the languages proposed by the W3C such as the RDF language. They include both the data and the schema describing them. However, this latter is a description of the entities in the dataset, but not a constraint on their properties. The schema can be defined partially, or even missing. Besides, the entities of a given class are not constrained by the structure of their class. Indeed, an entity belonging to a given class does not necessarily have all the properties defined for this class, and can even have some properties which are not defined in this class. Furthermore, two entities belonging to the same class do not necessarily have the same properties.

The nature of the RDF language offers a high flexibility when creating datasets. However, it makes the exploitation of these datasets difficult, as it is not obvious to understand their content.

Schema discovery approaches aim at providing a schema describing an RDF dataset, which can be useful for various data processing and data management tasks. Examples of such tasks are the followings:

Providing Applications with a Global View of an RDF Dataset. The discovered schema provides a summary of the classes corresponding to the entities in the dataset. This overview can be used to understand the content of an RDF dataset and to assess its fitness for the specific information requirements of a given application.

Interlinking RDF Datasets. One key feature of RDF datasets is that they include links to other datasets, which enables the navigation in the web of data. These links are represented by `owl:sameAs`[6] properties, and their determination is known as interlinking. Some tools have been proposed to perform this task, such as Knofuss[7] or Silk[8], which were used to link Yago [21] to DBpedia [3]. These tools require type and property information about the datasets in order to generate the appropriate `owl:sameAs` links between them. The discovered schema provides this information and could therefore be very useful for interlinking datasets.

Querying RDF Datasets. The lack of information about the classes, properties and resources contained in RDF datasets makes their interrogation difficult. Indeed, this information is required in order to formulate a query in the languages used for querying RDF datasets such as Sparql [30]. A schema describing the underlying structure of the data provides this information and such schema

---

[6] sameAs: https://www.w3.org/2001/sw/wiki/SameAs.

[7] Knofuss: https://technologies.kmi.open.ac.uk/knofuss.

[8] Silk: http://silkframework.org/.

would considerably ease query formulation. It could even be used to develop tools that assist user while formulating the queries, such as the one proposed in [8]. In addition, providing a schema describing a dataset allows the creation of an index over the entities to accelerate query answering. The schema could also enable the selection of the relevant sources while executing a query over a distributed dataset.

The above tasks are examples among many others to illustrate the usefulness of a schema describing an RDF dataset, and to show why schema discovery and understanding data have been identified as key challenges in data management [1].

3   Overview of the Approach

Our scalable schema discovery approach aims to extract a schema that captures the structure of the entities contained in a large RDF dataset, which cannot be managed by the existing approaches due to their complexity. The approach consists in extracting the implicit classes of the entities as well as the properties describing these classes.

In this section, we present some preliminary definitions used throughout the paper and we introduce the general principle of our proposal.

An RDF *dataset* D is a set of RDF(S)/OWL triples $D \subseteq (\mathcal{R} \cup \mathcal{B}) \times \mathcal{P} \times (\mathcal{R} \cup \mathcal{B} \cup \mathcal{L})$, where $\mathcal{R}$, $\mathcal{B}$, $\mathcal{P}$ and $\mathcal{L}$ represent resources, blank nodes (anonymous resources), properties and literals respectively. A dataset can be seen as a graph where vertices represent resources, blank nodes and literals, and where edges represent properties.

Example 1. Figure 1 presents an example of RDF dataset. The vertices represented as ovals are the resources, the ones represented as rectangles are literals. Each edge represents a property, and its label corresponds to the property name. For example, the resource e_1 is described by the following triples:

$$\langle e_1, id, 01 \rangle$$
$$\langle e_1, name, Ester \rangle$$
$$\langle e_1, authorOf, e_5 \rangle$$

In the sequel, for the sake of brevity, the properties *name, id, publish, gender, title, conference, year, rank* will be respectively replaced by $p_1, p_2, p_3, \ldots, p_8$.

In such a dataset, an *entity* e is either a resource or a blank node, that is, $e \in \mathcal{R} \cup \mathcal{B}$. We introduce a function denoted by $^-$ which returns the properties of an entity. It is defined as follows:

$$^- : \mathcal{R} \cup \mathcal{B} \to \mathcal{P}$$
$$e \quad \mapsto \{p \in \mathcal{P} \mid \langle e, p, o \rangle \in D\}$$

Example 2. The entities $e_1, e_2, \ldots, e_7$ extracted from the example of Fig. 1 are described as follows:
$\overline{e_1} = \{p_1, p_2, p_3\}$, $\overline{e_2} = \{p_1, p_2, p_3, p_4\}$, $\overline{e_3} = \{p_2, p_3, p_4\}$, $\overline{e_4} = \{p_2, p_5, p_6, p_7\}$, $\overline{e_5} = \{p_2, p_5, p_6\}$, $\overline{e_6} = \{p_1, p_2, p_5, p_8\}$, $\overline{e_7} = \{p_2, p_5, p_7\}$.

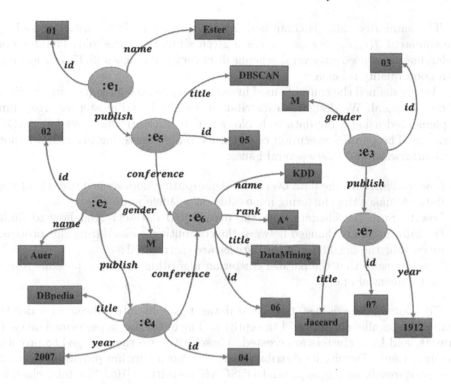

Fig. 1. An example of RDF dataset describing authors, publications and conferences

Similarly to the concept of class in data modeling, a class in an RDF dataset represents a set of individuals sharing some properties. The aim of our approach is to discover the implicit schema by grouping entities having similar structures, i.e. entities described by similar properties. The resulting groups represent the classes of the implicit schema describing the dataset.

Definition 1. *A schema S describing a dataset D is composed of a set of classes $\{C_1, \ldots, C_n\}$, where each C_i is described by a set of properties $\{p_1^i, \ldots, p_m^i\}$.*

The similarity between entities could be evaluated using any index that measures the similarity between finite sets such as *Sørensen-Dice index* [12], *Overlap indexes* [2] and *Jaccard Index* [16]. In our context, the properties describing the entities represent the finite sets. Two entities are similar if they share a number of properties which is equal to or higher than a given threshold. In our work, we evaluate the similarity between two entities e_i and e_j using the *Jaccard index*, which is defined as the size of the intersection of the property sets divided by the size of their union [16]:

$$J(e_i, e_j) = \frac{|\overline{e_i} \cap \overline{e_j}|}{|\overline{e_i} \cup \overline{e_j}|}$$

The similarity value is comprised between 0 and 1. Two entities e_i and e_j are similar if $J(e_i, e_j) \geq \epsilon$, where ϵ is a given similarity threshold. The *Jaccard index* has been used in several schema discovery approaches [9,17,18], leading to a good quality schema.

Having defined the concepts used in our paper, we now introduce an overview of our proposal. We designed a distributed density-based clustering algorithm implemented using a big data technology which efficiently manages large RDF datasets. The parallel execution of a density-based clustering algorithm is not straightforward and raises several issues:

- how to distribute the data over several computing nodes when the size of the dataset makes the clustering impossible on a single node?
- how to form the clusters from the distributed dataset? And how to limit the information exchanged between the computing nodes during this process, given that the neighbors of the entities are distributed?
- how to ensure that the parallel clustering algorithm provides the same result as a sequential one?

To address these issues, the initial dataset is split into subsets in order to enable the parallel clustering of the entities. The clustering is performed on each subsets, and local clusters are created. These latter are then merged to provide the final result. Despite its distributed design, our algorithm provides the same clustering result as the sequential DBSCAN algorithm [10]. The final clusters represent the classes of the schema describing the considered dataset.

Figure 2 gives an overview of our approach, focusing on the parallelization of the processes and the communications among the computing nodes.

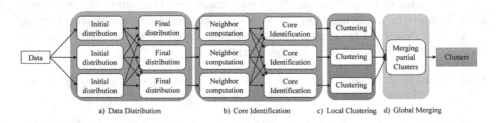

Fig. 2. Overview of our schema discovery approach

In the data distribution phase, chunks of entities are created according to the properties describing the data. Each chunk contains entities sharing some common properties; these entities are therefore likely to be similar. Our distribution method ensures that all the similar entities are grouped together in at least one chunk, such that two similar entities will be compared at least once. This way, all the relevant comparisons will be performed during the clustering of the chunks.

Once the chunks are created, the neighborhood of the entities is identified within each chunk. Then for each entity, the lists of its neighbors, which could be distributed over several chunks, are consolidated into one list by exchanging information between the computing nodes. During this stage, entities having dense neighborhoods, called *core entities*, are identified.

Based on the entities having a dense neighborhood, the local clusters are built in each chunk according to the density principle. To create a local cluster, we start with an arbitrary entity having a dense neighborhood and we retrieve all its similar entities. Then, their neighbors which have dense neighborhoods are retrieved and recursively added to the local cluster.

Finally, the clusters which have elements distributed over several chunks and belonging to distinct local clusters are built. These local clusters are merged to form the final clusters. Two clusters are merged if they share a core entity.

To achieve good performances, our proposal is implemented using Spark, an open source distributed computing framework with (mostly) in-memory data processing engine suitable for processing large datasets [29]. As it is always the case of distributed computing frameworks, the operations that need communications between nodes are costly. Some operations within Spark trigger an event known as a *shuffle*. The shuffle is Spark's mechanism for re-distributing data. It involves copying data across executors and machines, making it a complex and costly operation. As explained above, we have proposed a novel distribution method which both reduces communications between nodes and minimizes the need of Spark's shuffle operations.

The concepts and algorithms of our scalable schema discovery approach are detailed in the following sections.

4   Distributing Data over Computing Nodes

The distribution of data plays an important role in the parallelization of our algorithm. The initial dataset is first divided into chunks which could be clustered in parallel by the computing nodes. Our novel distribution principle ensures that there is no overhead communication between the computing processes, and that clustering a chunk does not require any data located in other chunks. As a consequence, we ensure that there are no useless data transfers between the computing nodes. The distribution method must ensure that enough information is provided to merge the clusters that span across several chunks; in our proposal, the replicated entities are used to perform the merging.

In this section, we first show how to split the initial dataset into chunks while meeting the above requirements. As the initial data distribution may create chunks having a size which exceeds the capacity of a calculating node, we then explain how to further decompose such large chunks.

4.1   Initial Distribution

The intuition behind our proposal is to group all similar entities sharing some common properties into chunks. Indeed, according to the similarity index, two entities are similar if they share a number of properties higher than a given threshold. Entities that could be similar are grouped together in at least one chunk, and will be compared during the computation of their neighborhood. Comparisons of entities inside each chunk will be performed later. If two given entities are not grouped together in any of the resulting chunks, this means that they are not similar.

A chunk of data is defined as follows:

Definition 2. *A chunk for a set of properties* $P \subseteq \mathcal{P}$ *denoted by* $[P]$ *is a subset of entities having the properties of P in their description:* $e \in [P] \implies P \subseteq \overline{e}$

Entities have to be distributed across several chunks to be efficiently clustered. We first describe a naive assignment of entities to chunks in order to give the idea behind the distribution principle. Then, an optimization is detailed.

The naive approach consists in assigning the entities according to all the properties describing them. An entity e described by the properties $\overline{e} = \{p_1, p_2, \ldots, p_n\}$ will be assigned to the chunks $[p_1], [p_2], \ldots, [p_n]$. In other words, e is grouped with all the entities that share at least one property with e.

Definition 3. *With the* Naive Assignment, *each entity is assigned to the chunks for each of its properties:*

$$\forall e, \forall p \in \overline{e}, e \text{ is assigned to } [p].$$

Proposition 1. *(*Naive Assignment *Soundness). With the* Naive Assignment, *two similar entities will be grouped into at least one common chunk, i.e. all required comparisons will be performed at least once.*

Proof. According to our similarity index, two similar entities must have at least one property in common. Using the *Naive Assignment*, they will be assigned to at least one common chunk.

The *Naive Assignment* suffers from an important drawback. Two similar entities could be grouped redundantly many times. For example, the entities $\overline{e_1} = \{p_1, p_2, p_3\}$ and $\overline{e_2} = \{p_1, p_2, p_3, p_4\}$ will be both assigned to the chunks $[p_1], [p_2], [p_3]$ and consequently, they will be compared three times.

In our approach, we do not consider all the properties while assigning the entities to the chunks to limit the number of duplications and reduce the cost of the comparison process. To this end, we introduce the notion of *dissimilarity threshold*, which represents the number of properties to consider in order to decide whether this entity could be similar to any other one. The assignment is defined in two steps. Firstly, we calculate for each entity its *dissimilarity threshold*, which allows to choose the number of chunks an entity has to be assigned to. Secondly, we assume that a total order relation is defined on the properties; the chunks to which the entities are assigned are chosen according to this order.

For example, let us consider $\overline{e_2} = \{p_1, p_2, p_3, p_4\}$ and $\epsilon = 0.7$. If e_2 differs from any other entity by more than two properties, the other entity can not be similar to e_2. For instance, an entity $\overline{e'} = \{p_3, p_4, p_5\}$ will not be similar to e_2 because $\overline{e_2} \setminus \overline{e'} = \{p_1, p_2\}$ has two elements. We will show that it is sufficient to assign e_2 to the chunks $[p_1]$ and $[p_2]$ to ensure that all its similar entities are within these chunks. The entities which are not assigned to these chunks can not be similar to e_2.

However, properties can not be selected randomly, otherwise, this will prevent similar entities to be grouped in the same chunks and compared later. For example, let us consider the similar entities e_2 and e_3 where $\overline{e_2} = \{p_1, p_2, p_3, p_4\}$ and $\overline{e_3} = \{p_2, p_3, p_4\}$. Assuming that the similarity threshold is $\epsilon = 0.7$, and considering the dissimilarity threshold, the entity e_2 can be assigned to $[p_1]$, $[p_2]$ and e_3 only to $[p_3]$. e_2 and e_3 are not grouped in a chunk even though they are similar. We can see that randomly assigning these entities does not guarantee that they are compared even if they are similar. This problem can be solved by defining a total order on the properties and selecting the properties according to this order. By assigning the entities according to an order in this example, the entity e_3 would be assigned to $[p_2]$ instead of $[p_3]$. Therefore, e_2 and e_3 would be grouped in the chunk $[p_2]$ and compared during the computation of their neighborhood.

We will now formalize these intuitions. Let us introduce a proposition, which expresses that if the properties of two entities differ to a certain extent, these entities can not be similar.

Proposition 2. *Let e_1 and e_2 be two entities. If $|\overline{e_1} \setminus \overline{e_2}| \geq |\overline{e_1}| - \lceil \epsilon \times |\overline{e_1}| \rceil + 1$ then e_1 and e_2 can not be similar.*

Proof. Suppose that $|\overline{e_1} \setminus \overline{e_2}| \geq |\overline{e_1}| - \lceil \epsilon \times |\overline{e_1}| \rceil + 1$. We have $|\overline{e_1} \setminus \overline{e_2}| = |\overline{e_1}| - |\overline{e_1} \cap \overline{e_2}|$. Thus, $|\overline{e_1}| - |\overline{e_1} \cap \overline{e_2}| \geq |\overline{e_1}| - \lceil \epsilon \times |\overline{e_1}| \rceil + 1$. By eliminating $|\overline{e_1}|$ on both sides, we obtain $|\overline{e_1} \cap \overline{e_2}| \leq \lceil \epsilon \times |\overline{e_1}| \rceil - 1$ which implies that $|\overline{e_1} \cap \overline{e_2}| < \lceil \epsilon \times |\overline{e_1}| \rceil$. According to the definition of the Jaccard similarity index, this formula implies that e_1 and e_2 can not be similar.

We now define the notion of *dissimilarity threshold* for an entity e. Note that the *dissimilarity threshold* as defined in our work is based on the Jaccard similarity index. Using another index would require to propose another definition of this threshold based on this index.

Definition 4. *The* dissimilarity threshold *for an entity e is the number* $dt(e) = |\overline{e}| - \lceil \epsilon \times |\overline{e}| \rceil + 1$.

The following definition presents the optimized assignment.

Definition 5. *Let $<_\mathcal{P}$ be a total order on the properties describing a dataset, and let e be an entity with $\overline{e} = \{p_1, p_2, \ldots, p_n\}$ and $p_i <_\mathcal{P} p_{i+1}$ for $1 \leq i < n$. With the* optimized assignment, *an entity e is assigned to the chunks $[p_1]$, $[p_2]$, $\ldots$, $[p_{dt(e)}]$. We denote by $ch(e)$ the set of properties $\{p_1, p_2, \ldots, p_{dt(e)}\}$.*

Proposition 3. *With the* optimized assignment, *all the comparisons required for the clustering will be performed at least once.*

Proof. We have to show that if two entities are similar, they are both assigned to at least one common chunk. Let e_1 and e_2 be two similar entities. We have $|\overline{e_1} \cap \overline{e_2}| \div |\overline{e_1} \cup \overline{e_2}| \geq \epsilon$. Thus, $|\overline{e_1} \cap \overline{e_2}| \geq \epsilon \times |\overline{e_1} \cup \overline{e_2}|$ which implies that $|\overline{e_1} \cap \overline{e_2}| \geq \lceil \epsilon \times |\overline{e_1}| \rceil$ and $|\overline{e_1} \cap \overline{e_2}| \geq \lceil \epsilon \times |\overline{e_2}| \rceil$. This implies that $|\overline{e_1}| - |\overline{e_1} \cap \overline{e_2}| \leq |\overline{e_1}| - \lceil \epsilon \times |\overline{e_1}| \rceil$. As $|\overline{e_1} \setminus \overline{e_2}| = |\overline{e_1}| - |\overline{e_1} \cap \overline{e_2}|$, we obtain $|\overline{e_1} \setminus \overline{e_2}| \leq |\overline{e_1}| - \lceil \epsilon \times |\overline{e_1}| \rceil$. As $|ch(e_1)| = dt(e_1) > |\overline{e_1}| - \lceil \epsilon \times |\overline{e_1}| \rceil$, we have $ch(e_1) \cap \overline{e_2} \neq \emptyset$.

We can show likewise that $ch(e_2) \cap \overline{e_1} \neq \emptyset$. Consequently, $ch(e_1)$ and $ch(e_2)$ contain both an element of $\overline{e_1} \cap \overline{e_2}$.

If there is a total order on the set of properties, we can choose the infimum of $\overline{e_1} \cap \overline{e_2}$ for $ch(e_1)$ and $ch(e_2)$. In this case, $ch(e_1) \cap ch(e_2) \neq \emptyset$. This means that at least one chunk will contain both e_1 and e_2.

In our work, we propose to order the properties according to their selectivity. The selectivity of a property is one minus the ratio of the number of entities described by this property, over the total number of entities. A high selectivity means that few entities are described by the property. In our approach, the properties are ordered from the most to the least selective. This will lead to chunks that are less dense. More meaningless comparisons will then be skipped and the clustering of each chunk will be more efficient.

Example 3. Let us consider a dataset D described by the set of properties $\mathcal{P} = \{p_i \mid i \in [1,8]\}$ and containing the set of entities $\{e_i \mid i \in [1,7]\}$ where each entity is described by:
$\overline{e_1} = \{p_1, p_2, p_3\}$, $\overline{e_2} = \{p_1, p_2, p_3, p_4\}$, $\overline{e_3} = \{p_2, p_3, p_4\}$, $\overline{e_4} = \{p_2, p_5, p_6, p_7\}$, $\overline{e_5} = \{p_2, p_5, p_6\}$, $\overline{e_6} = \{p_1, p_2, p_5, p_8\}$, $\overline{e_7} = \{p_2, p_5, p_7\}$.

In our example, the similarity threshold is set to $\epsilon = 0.7$. With respect to their selectivity, the order on the properties is $p_8 <_{\mathcal{P}} p_4 <_{\mathcal{P}} p_6 <_{\mathcal{P}} p_7 <_{\mathcal{P}} p_1 <_{\mathcal{P}} p_3 <_{\mathcal{P}} p_5 <_{\mathcal{P}} p_2$. Distributing the entities over the chunks with the *optimized assignment* provides the result presented in Fig. 3.

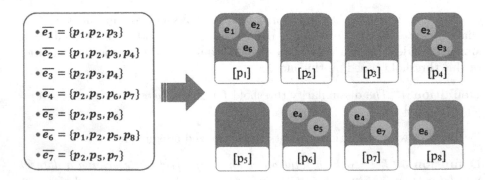

Fig. 3. Distributing the dataset D over data chunks

For example, the dissimilarity index of the entity e_2 is equal to $dt(e_2) = 4 - \lceil 0.7 \times 4 \rceil + 1 = 2$. The two most selective properties describing e_2 are p_1 and

p_4, e_2 is therefore assigned to $[p_1]$ and $[p_4]$. This assignment ensures that e_2 is grouped with each of its neighbors at least once, and therefore will be compared to each of them at least once (e_2 is grouped with its neighbors e_1 and e_3 in chunks $[p_1]$ and $[p_4]$ respectively).

Both the empty chunks and the ones containing a single entity such as $[p_2]$ and $[p_8]$ are deleted.

Algorithm 1 formalizes the data distribution stage. It requires the similarity threshold ϵ, used to compute the dissimilarity threshold, and to define the chunks $ch(e)$ for each entity e.

Algorithm 1. Distributing Entities

Input: the dataset D, the similarity threshold ϵ
1: **for all** entity e in D **do in parallel**
2:     **for all** property p in $ch(e)$ **do**
3:         $[p] = [p] \cup \{e\}$
4:     **end for**
5: **end for**
6: Merge the chunks generated by the parallel execution for the same properties
7: **return** the chunks

The computation of the assignment of each entity (line 1–5) is performed in parallel on the computing nodes. The partial chunks are then merged to obtain the final chunks.

The distribution process may result in some chunks which are too large to be clustered by a single node. This will require a further partitioning, described in the following section.

4.2   Managing Big Chunks

Since a chunk $[p]$ contains a set of entities described by the property p, the number of entities within $[p]$ could exceed the computing capacity of a single node which prevents the execution of the clustering. In that case, each large chunk $[p]$ is further divided according to other properties.

We introduce the *capacity* parameter which determines whether a chunk is exceeding the computing capacity of a single node.

In the case of a large chunk $[p]$ that contains a number of entities higher than *capacity*, the algorithm creates sub-chunks for each property describing the entities within $[p]$ except p, then assigns each entity in $[p]$ to a sub-chunk if it is described by the property used to generate the sub-chunk:

$$\forall e \in [p], \forall p_i \in \overline{e}, [\{p, p_i\}] = [\{p, p_i\}] \cup \{e\}$$

Recursively, the size of all the resulting chunks is evaluated and those exceeding the capacity of a node are divided until all the chunks have a number of entities lower than the computation capacity of a single node.

At the end of this process, chunks of the initial dataset are created, all of them having a number of entities that could be efficiently clustered by a single node. The distribution of entities over chunks does not require any information sharing between the nodes.

Example 4. For example, if the capacity of a node is 3 and if we consider the chunk $[p_2] = \{e_1, e_2, e_3, e_4, e_5\}$ of the previous example, its size is greater than the capacity. $[p_2]$ will be further divided into sub-chunks, for example $[p_2, p_1] = \{e_2\}$ and $[p_2, p_3] = \{e_1, e_2\}$.

Algorithm 2 evaluates the size of each chunk and divides those exceeding the *capacity*. This method is applied recursively until the size of all the chunks is lower than the *capacity* parameter.

Algorithm 2. Splitting Big Chunks

Input: chs: the chunks, cap: the capacity of computing nodes
1: **for all** $[P] \in chs \mid |[P]| > cap$ **do in parallel**
2:    **for all** $e \in [P]$ **do**
3:       **for all** $p_i \in \bar{e} \setminus P$ **do**
4:          $[P \cup \{p_i\}] = [P \cup \{p_i\}] \cup \{e\}$
5:       **end for**
6:    **end for**
7: **end for**
8: Merge the chunks generated by the parallel execution for the same properties
9: **return** the chunks

Once the chunks have been generated, the computation of the entities neighborhoods will be performed on each of them. This process is described in the following section.

5   Core Identification

In a clustering algorithm, data points which are close to each other are grouped together. Our approach is density-based and the notion of "closeness" is related to the one of density of an entity's neighborhood. In order to form a cluster from a given entity, the neighborhood of this entity has to contain a sufficient number of points; in other words, the density of its neighborhood has to exceed a given density threshold. This section describes the identification of entities having a dense neighborhood.

Let us first recall some definitions used by the DBSCAN algorithm [10].

Definition 6. *The ϵ-neighborhood of an entity e is the set of entities which are similar to e with a threshold of ϵ.*

$$neighborhood_\epsilon(e) = \{e_i \in D \mid J(e, e_i) \geq \epsilon\}$$

According to the ϵ-neighborhood of the entities, three kinds of points are distinguished: *core entities* with at least $minPts$ entities in their ϵ-neighborhood, *border entities*, which are not core entities but have at least one core entity in their ϵ-neighborhood, and *noise entities* which have no core entity in their ϵ-neighborhood. Noise points are not assigned to a cluster.

Definition 7. *An entity e is a core entity if the number of entities within its ϵ-neighborhood is greater than the density threshold $minPts$, i.e. $|neighborhood_\epsilon(e)| \geq minPts$.*

Once the ϵ-neighborhood is computed for each entity, the core entities are identified. However, as the data is partitioned in chunks in our approach, the neighborhood of entities may span across several chunks. In such case, the number of neighbors of each entity can not be computed only from one chunk.

Example 5. If we set $minPts$ to 2 in our example, the entity e_2 that has e_1 and e_3 in its neighborhood is a core entity. But after the assignment to the chunks, the neighborhood of e_2 is distributed over the chunks p_1 and p_4. If the comparisons between entities are done within each chunk independently, the number of e_2's neighbors in each chunk does not exceed $minPts$ and e_2 is not considered as a core.

In our approach, core identification is a two-stage process, as illustrated by Fig. 2b.

In the first step, the ϵ-neighborhood of each entity is calculated in parallel within each chunk. Calculating the ϵ-neighborhood of the entities represents the most expensive operation in a density-based clustering algorithm since it requires comparing all the possible pairs of entities. Our algorithm operates on chunks containing a number of entities small enough to allow a fast execution and to skip a number of meaningless comparisons. Moreover, this operation is parallelized over the calculating nodes to provide the best performances. In the second step, the neighbors discovered in each chunk are grouped by entity, and the list of the corresponding neighbors of each entity in the whole dataset is built. The core entities are the ones having a number of neighbors greater or equal to $minPts$.

Example 6. With $minPts = 2$, the cores identified in Example 3 are e_2 and e_4. For example, the algorithm finds that the neighbors of e_2 are e_1 and e_3 respectively belonging to the chunks $[p_1]$ and $[p_4]$. Then, these lists are merged to provide the complete list of e_2's neighbors: $neighborhood_\epsilon(e_2) = \{e_1, e_3\}$. Finally, e_2 is identified as a core entity because the number of entities within its neighborhood is equal to $minPts$.

Algorithm 3 describes the core identification stage, executed in parallel withing each chunk.

This algorithm provides the list of neighbors of each entity in each chunk (lines 1–5) and then merges the lists (line 6). The lists of neighbors for each entity are exchanged between the calculating nodes in order to group each entity

---

Algorithm 3. Core Identification

Input: chs: the chunks, ϵ: the similarity threshold, $minPts$: the density threshold
 1: **for all** $[P] \in chs$ **do in parallel**
 2:     **for all** $e \in [P]$ **do**
 3:         $neighborhood_\epsilon(e) = \{e_i \in [P] \mid J(e, e_i) \geq \epsilon\}$
 4:     **end for**
 5: **end for**
 6: Merge the local neighborhoods to compute the complete list of neighbors of each entity
 7: **for all** $e \in D$ **do**
 8:     **if** $|neighborhood_\epsilon(e)| \geq minPts$ **then**
 9:         $cores = cores \cup \{e\}$
10:     **end if**
11: **end for**
12: **return** cores

---

with all its neighbors. Then, the algorithm tags the entities having a number of neighbors greater than or equal to $minPts$ as core entities (line 7–11).

Having computed the neighborhood of each entity and identified the core entities, the clustering is performed locally in each chunk. This process is described in the following section.

6   Local Clustering

During the local clustering, clusters are computed in each chunk. A local cluster contains entities which are similar inside a chunk.

The clustering stage is executed in parallel in the different chunks independently; the distribution strategy ensures that the clustering within a chunk does not require any data from any other chunk. This minimizes the costly overhead communications between the chunks and speeds up the clustering stage.

In a density-based clustering algorithm, the clusters are built according to the density-reachable principle, introduced by the DBSCAN algorithm [10]. The corresponding definitions are presented hereafter.

Definition 8. *An entity e is* directly density-reachable *from an entity e' wrt. ϵ and minPts if and only if e' is a core entity and e is in its ϵ-neighborhood, i.e. $|neighborhood_\epsilon(e')| \geq minPts$ and $e \in neighborhood_\epsilon(e')$.*

Definition 9. *An entity e is* density-reachable *from an entity e' wrt. ϵ and minPts if there is a chain of entities $e_1, \ldots, e_z$, $e_1 = e'$, $e_z = e$ such that e_{i+1} is directly density-reachable from $e_i, \forall i \in \{1, \ldots, z\}$.*

The clusters are built based on the core entities. As the neighborhood of entities have been computed and the core entities identified, all the required information is available to generate the clusters locally in each chunk. Only core entities will generate clusters by adding their neighbors as elements of the clusters. Other entities will be either borders in some core's neighborhood, or noise entities.

For each core entity e, a cluster C containing e and its neighbors is created. The core entities within the ϵ-neighborhood of e are then retrieved and their neighbors are added to the cluster C. The neighbors of the cores in C are recursively added to the cluster until the expansion stops on border entities.

Figure 2c shows the parallelization of this operation; the clustering is performed on each chunk independently from the others and provides a local clustering result.

Example 7. Clustering the chunks obtained in Example 3 based on the cores identified in Example 6 provides the result presented in Fig. 4. The clusters are denoted by the ids of the chunks followed by an index. In our example, four local clusters are built, $c_{p_1.1}$, $c_{p_4.1}$, $c_{p_6.1}$ and $c_{p_7.1}$ respectively within the chunks p_1, p_4, p_6 and p_7.

The core entity e_2 in the chunks $[p_1]$ and $[p_4]$ forms a cluster within each chunk by grouping all the entities that are density-reachable from e_2. The same principle is applied for all the core entities in the other chunks. To prevent ambiguity, the clusters are annotated by the ids of the chunks followed by an index. An entity which do not belong to any cluster, such as e_6, could be assigned to a cluster during the merging stage if it belongs to a cluster in another chunk, or could remain a noise entity.

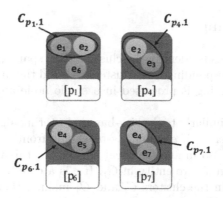

Fig. 4. Building local clusters in each chunk

Algorithm 4 computes the clusters in every chunk generated in the previous stage. It iterates over the core entities previously identified and creates for each one a cluster containing the core entity and its neighbors (line 6). The algorithm then checks among the added neighbors those which are cores, and adds their neighbors to the cluster (lines 7–9). The algorithm recursively adds the neighbors of the cores to the current cluster until all its cores are checked and the expansion stops on border entities. The same operation is repeated with another core which has not been visited yet, until all the cores are clustered. The final output of the algorithm is the set of local clusters.

Algorithm 4. Local Clustering

Input: chs: the chunks, cores: the core entities
1: **for all** $[P] \in chs$ **do in parallel**
2:   is-visited $= \emptyset$
3:   **for all** $e \in [P]$ **do**
4:     **if** $e \in cores$ and $e \notin$ is-visited **then**
5:       is-visited $=$ isVisited $\cup \{e\}$
6:       Create a new cluster $C = \{e\} \cup neighborhood_\epsilon(e)$
7:       **for all** $e' \in C \mid e' \in cores$ and $e' \notin$ is-visited **do**
8:         $C = C \cup \{e'\} \cup neighborhood_\epsilon(e')$
9:       **end for**
10:     **end if**
11:     local-clusters $=$ local-clusters $\cup\, C$
12:   **end for**
13: **end for**
14: **return** local-clusters

In the next section, we will show how to build the final clusters from the local ones.

7   Global Merging

The merging stage aims to identify the clusters than span across several chunks, and to merge the corresponding local clusters to build the final result. As we can see in Fig. 2, the merging is processed in a single node and provides the final clustering result.

In our approach, similarly to density-based clustering algorithms, an entity e is assigned to a cluster C_i if e is density-reachable from a core entity in C_i. If this same entity e is also in another local cluster C_j, this means that e is also density-reachable from a core entity in C_j. If e is a core, it represents a bridge between the entities in the clusters C_i and C_j making them density-reachable from one another.

Figure 5 gives an overview of this principle; core entities are represented in orange and border entities in green. As shown in this figure, the entities within the clusters C_1 and C_2 are density-reachable from the common core entity e_i, which makes all of them density-reachable. Therefore, these entities should be assigned to the same cluster. In that case, the local clusters are merged.

The merging stage identifies the clusters than span across different chunks by finding the local clusters that share a common core entity and by merging them. If a border entity is assigned to different clusters during the clustering stage, it would be randomly assigned to one of these clusters during the merging stage.

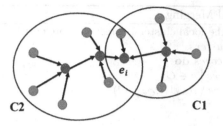

Fig. 5. An illustration of the cluster merging principle (Color figure online)

All the entities which are not assigned to a cluster are considered as noise. This process provides the final clusters, ensuring that the same clusters as DBSCAN are generated.

Example 8. Figure 6 presents the final clusters obtained by merging the local clusters of Example 7. For instance, the clusters $c_{p_1.1}$ and $c_{p_4.1}$ are merged since they share a common core entity e_2. The resulting final clusters represent the classes of the schema. The properties of these classes are the union of the properties describing the entities within a cluster ($Class_1 = \{p_1, p_2, p_3, p_4\}$ and $Class_2 = \{p_2, p_5, p_6, p_7\}$). Noise entities such as e_6 are considered as not representative enough to generate a class in the extracted schema.

This descriptive schema shows that the RDF dataset contains instances of the class *author* described by the set of properties $\{publish, id, name, grade\}$ and the class *publication* described by the properties $\{id, title, conference, year\}$.

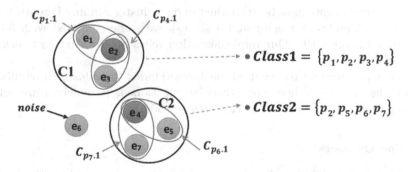

Fig. 6. The final clusters corresponding to the classes of the discovered schema for the dataset D

Algorithm 5 describes the cluster merging process. Two clusters are merged if they share a core entity. The merging algorithm therefore iterates over core entities (lines 2–6). For each core, clusters containing this core are identified (line 3) then merged (line 5). This final step is executed on one computing node and is not parallelized.

Algorithm 5. Global Merging

Input: localClusters: the local clusters, cores: the core entities
1: $clusters \leftarrow localClusters$
2: **for all** $e \in D \mid e \in cores$ **do**
3:     $lc_e = \{C \in clusters \mid e \in C\}$
4:     $clusters = clusters \setminus lc_e \cup (\cup_{C \in lc_e} C)$
5: **end for**
6: **return** clusters

---

8 Experiments

This section presents our experiments to show the effectiveness of our approach both in terms of quality and runtime.

We have first evaluated the quality of the discovered schema. We have considered a dataset including type definitions and we have used them as the ground truth. We have compared the discovered classes with those provided by the dataset, and we have computed the precision and the recall for each discovered class.

We have evaluated the scalability by showing the capacity of our algorithm to cluster large RDF datasets and studying its behavior on various datasets.

We have measured the algorithm *Speed-Up* to show the execution time improvement when increasing the number of computing nodes. We have also studied the efficiency when applied to real datasets.

Finally, we have compared the performances of our approach to the ones of NG-DBSCAN, an existing density-based clustering algorithm also implemented using Spark.

All the experiments have been conducted on a cluster running Ubuntu Linux consisting of 5 nodes (1 master and 4 slaves), each one equipped with 30 GB of RAM, a 12-core CPU. Our implementation relies on the Apache Spark 2.0 framework.

In our experiments, we have used the Jaccard index to evaluate the similarity between the entities. Where not otherwise mentioned, parameters are set as follows: ϵ to 0.8, $minPts$ to 3 and $capacity$ to 9000.

8.1 The Datasets

To evaluate the scalability of our approach, we have first used synthetic data generated using "IBM Quest Synthetic Data Generator" [15]. This well known generator was heavily used in the data mining community to evaluate the performances of frequent itemset mining algorithms. In our context, the generator produces the properties of each entity that will be used in our experiments, and allows to tune their characteristics.

The variable characteristics of the data considered in our experiments are (i) the size of the dataset to study the scalability of our algorithm, (ii) the total number of properties describing the dataset and (iii) the average number of dimensions (properties) of the entities.

Beside synthetic data, we have used real RDF datasets of different sizes extracted from DBpedia[9]. DBpedia is a project aiming to extract structured content from the information created in the Wikipedia project and to make it available on the web. DBpedia allows users to semantically query relationships and properties of Wikipedia resources, including links to other related datasets. DBpedia is split into different subsets according to the language used.

In our evaluations, we have extracted from DBpedia subsets of patterns which represent all the existing combinations of properties describing the entities in the dataset. A pattern represents a combination of properties for which there is at least one instance in the dataset. Entities having exactly the same property sets are represented by a single pattern. To extract the patterns, we have used the approach proposed in [7]. Considering patterns instead of entities reduces the size of the input data and helps speeding up the clustering.

We have used DBpediaEn (1.23 million patterns), DBpediaFr (626 381 patterns), DBpediaEs (529 434 patterns), DBpediaNl (268 603 patterns), DBpediaUk (129 762 patterns) and DBpediaAr (63 000 patterns).

We have extracted from DBpedia the entities for which a type (class) has been defined, and we have considered them as the ground truth for evaluating the quality of the schema. In our evaluations, we have considered the entities having the following types: Aircraft, Artist, Athlete, Book, Disease, Newspaper, Region and TelevisionStation. These entities represent a reference to which the generated clusters are compared.

8.2 Evaluation of the Schema Quality

We have clustered the entities within DBpedia using our algorithm without considering the types of the entities. We have set $MinPts$ to 1, as we consider that at least two entities sharing similar properties are required to form a class. We have run our algorithms with several values of ϵ, ranging between 0.5 and 0.7. In the context of RDF datasets, ϵ represents the threshold ratio of shared properties required for two entities to be considered as neighbors.

The discovered classes are annotated with the most frequent type label associated to its entities.

Finally, we have evaluated the precision and the recall for each class. In our work, the precision and the recall are evaluated based on the comparison of the classes generated by our approach for the entities to the types of these entities as declared in the initial dataset. We have evaluated for each class both the precision and the recall. Each of the bar charts a, b and c of Fig. 7 shows, for a specific value of ϵ, both the precision and the recall.

The results presented in Fig. 7 show that our approach is able to detect all the considered classes of the entities within the dataset with good precision

---

[9] http://downloads.dbpedia.org/3.9/.

and recall when the value of ϵ is well defined (Fig. 7b). The recall of the class *Aircraft* is lower because the entities having this type are very heterogeneous.

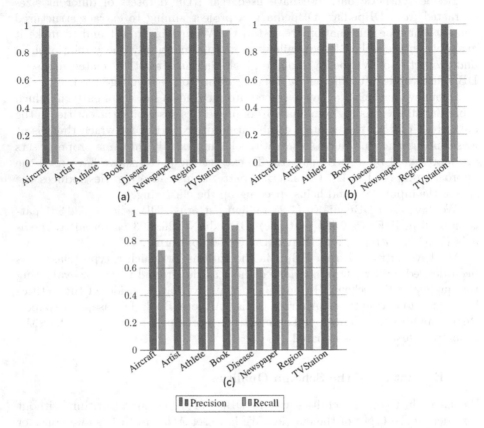

Fig. 7. Quality of the extracted classes for different values of ϵ ($\epsilon = 0.5$ (a), $\epsilon = 0.65$ (b), $\epsilon = 0.7$ (c))

In some cases, the entities within different classes can be described by similar property sets, they are therefore merged in a more general class. For instance, the classes *Artist* and *Athlete* were grouped into a more general class *Person*, as shown in Fig. 7a. For a higher value of ϵ (Fig. 7b), a higher number of shared properties is required for two entities to be considered as similar and the classes *Artist* and *Athlete* are both generated. When the value of ϵ is higher, the recall of some types decreases (Fig. 7c). As these types contain heterogeneous entities described by different properties, they were not considered as similar and therefore not grouped into the same cluster. A higher value of ϵ makes the algorithm more sensitive to small differences which can lead to similar entities assigned to different clusters and decrease the quality of the schema.

To conclude the experiments on the quality of the resulting classes, recall that clustering a dataset using our approach provides the same result as using the sequential DBSCAN algorithm. Previous works have shown that extracting a schema from an RDF dataset using DBSCAN provides a good quality result, with good precision and recall, and detects classes which were not declared in the dataset [18]. These results are in line with the ones provided in this section.

The following sections are devoted to our experiments for evaluating the performances of our approach when applied to large datasets, which is the main focus of the present paper.

8.3   Scalability

We have first evaluated the scalability or our approach using several synthetic datasets of different sizes. Additionally, we have studied the behavior of our algorithms on datasets with different characteristics: (i) datasets containing entities of different dimensions (10, 20, 30 and 40 properties per entity) and (ii) datasets where entities are described by different numbers of properties. We have also evaluated the speed-up of our approach with different configurations of the computing cluster, i.e. for different numbers of worker nodes. Finally, we have applied our algorithm on real datasets to illustrate its performances.

Figure 8 shows the algorithm runtime as a function of the dataset size for datasets having in average 10, 20, 30, and 40 properties in the description of their entities.

The results show the effectiveness of our algorithm to cluster large datasets, as it is able to cluster a dataset containing more than 5 million entities in 18 min, for a dataset containing entities described by an average of 10 properties.

The results are explained by the fact that during the distribution stage, chunks that contain a number of entities which does not exceed the calculating capacity of the cluster's nodes are created. Thus, each node executes clustering tasks by computing the similarity on a number of entities which does not require a high execution time. In addition, some meaningless comparisons are avoided while determining the neighborhood of each entity, since entities are compared only if they are grouped in the same chunk. Each node calculates efficiently the ϵ-neighborhood of the entities and the partial clusters in each chunk. Furthermore, the computations are distributed over the nodes of the clusters to minimize the communications overhead between the nodes, i.e. by avoiding the costly Spark's shuffle operations.

When the size of the dataset increases, the process requires more time. As the distribution stage produces a high number of chunks, each calculating node has to manage many more chunks. In addition, the chunks contain a higher number of entities and can be split recursively to generate chunks having a size which is lower than the capacity of the calculating nodes. This drop in the performance is more visible when the calculation's limits of the cluster are reached. This limit is reached at different levels according to the characteristics of the datasets as we can see in Fig. 8.

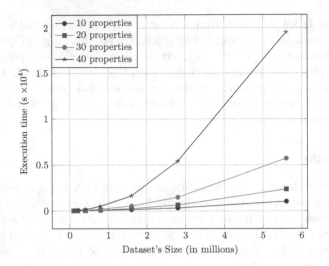

Fig. 8. Evaluating the scalability of our approach on different synthetic datasets

The same happens when the number of dimensions (i.e. properties) of the entities increase: this increases the number of entities within the chunks as the entities are distributed according to the properties, and also increases the number of chunks. We observe that the curves have the same behavior, but the limit is reached for different dataset's size. The limit is reached for a size of 5.8 M entities for datasets where entities are described by 10, 20 and 30 properties, while it is reached for a size of 2.8 M for datasets where entities are described by 40 properties.

We have studied the impact of the total number of properties describing the dataset. Figure 9 shows the execution time for datasets described by a number of properties that varies between 10k and 80k.

The experiments show that when the number of properties increases, the execution time decreases. Having a higher number of properties implies generating more chunks and getting a better distribution of the entities. This also produces smaller chunks, which do not require further partitioning. This accelerates the distribution and the clustering stages.

We have also studied the speed-up and the impact of the number of worker nodes in the Spark cluster on its execution time. These evaluations were conducted on a cluster composed of 1 master equipped with 4 GB of RAM, 4 core CPU. The number of workers varies from 2 to 8 and each worker is equipped with 16 GB of RAM and 6 cores CPU.

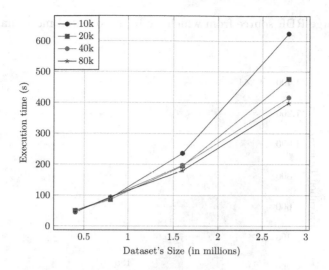

Fig. 9. Evaluating the impact of the number of properties on the execution time

Figure 10 shows the algorithm's speed-up as the number of worker nodes varies, considering datasets of a size between 500 000 and 3 000 000 entities.

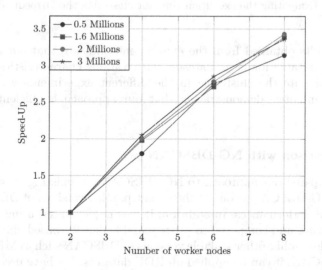

Fig. 10. Evaluating the speed-up for different cluster configurations

The experiments show that better performances and faster clustering are obtained when adding more worker nodes to the computing cluster. The obtained results demonstrate that our algorithm is scalable despite the size of the datasets.

Finally, we have evaluated the efficiency of our approach on real datasets. Figure 11 shows the ability to cluster real datasets, such as DBpedia English which is a large RDF source from which we have extracted more than 1 million patterns.

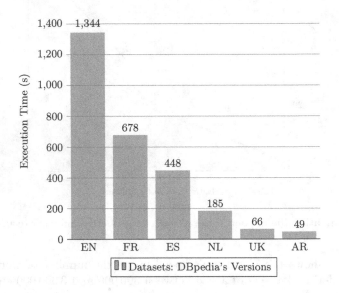

Fig. 11. Evaluating the execution time for clustering the DBpedia dataset

These results obtained from the experiments indicate that our approach is scalable and suitable for large datasets with various characteristics. The time needed to compute the clustering in the different experiments was always in the order of minutes, demonstrating that our approach is efficient in several scenarios.

8.4 Comparison with NG-DBSCAN

We have compared our approach to NG-DBSCAN, an existing clustering algorithm [19]. NG-DBSCAN is one of the recent parallel versions of DBSCAN that provides good performances. In addition, it was implemented using the Apache Spark framework and compared to other scalable density-based clustering algorithms. Besides, unlike other scalable versions of DBSCAN such as MR-DBSCAN and RP-DBSCAN, it can be applied on RDF datasets. We have used the source code provided by the authors and available online[10].

Figure 12 presents the logarithmic function of the execution time needed by both algorithms to cluster datasets of different sizes. We use the logarithmic scale to represent the execution time because the gap between the performances of the two algorithms is important and it prevents us from comparing their behaviours.

---

[10] https://github.com/alessandrolulli/gdbscan.

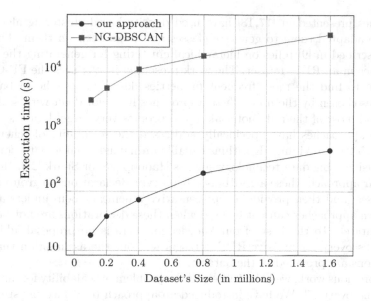

Fig. 12. Comparing our clustering algorithm with NG-DBSCAN

Our results show that both curves have a similar shape, and that our approach always outperforms NG-DBSCAN. This is due to the fact that the implementation of NG-DBSCAN applies many shuffle operations, which increases the communication cost and therefore, the execution time of the algorithm. On the other hand, our algorithm smartly distributes the data so as to reduce the cost of communication between the worker nodes during the computation of the clusters, thus considerably reducing the execution time.

As a conclusion, the results obtained throughout the different experiments shown that our proposal performs well both in term of quality of the generated classes and the runtime speed of the generation process. It allows performing fast density-based clustering on large synthetic or real datasets and provides a good quality result, with good precision and recall of the detected classes describing the dataset. Moreover, our algorithm speeds up and provides better performances when more computing nodes are added to the Spark cluster which makes it scalable to very large datasets. In addition, unlike the existing scalable implementations of DBSCAN, it provides the same clustering result as the one which would be generated by the sequential DBSCAN algorithm. Finally, it outperforms NG-DBSCAN, a recent density-based clustering algorithm that provides good performances.

9 Related Work

Several approaches have been proposed for schema discovery in RDF datasets. Some of these approaches have used clustering algorithms to group similar entities in order to form the classes representing the schema. Among these works, the approaches presented in [17,18] have used density-based clustering algorithms and have adapted them to generate classes and links between them. The approach described in [9] relies on hierarchical clustering for generating the underlying types in an RDF dataset. The work presented in [27] uses the FP-Growth algorithm to find the most frequent properties describing a schema based on the classes chosen by the user. These approaches have not dealt with scalability issues, and most of them do not scale up to process very large datasets.

Some approaches have specifically addressed the scalability of schema discovery [4,5,24], providing algorithms capable of managing large datasets implemented using a big data technology such as Hadoop [28] or Spark [29]. However, unlike our approach, these algorithms rely on type declarations to group entities into classes, and then provide a representative schema to help understand the data. Such approaches can not be used when these declarations are not provided in the dataset. To the best of our knowledge, there is no proposal addressing schema discovery for massive RDF datasets without the assumption that type declarations are provided in the dataset.

In a previous work, we have addressed the problem of scalability for automatic schema discovery [7]. We have introduced an approach to reduce the size of the input RDF dataset by building a condensed representation composed of all the existing combinations of properties in the dataset. The clustering is performed on the condensed representation instead of the initial dataset. However, in the case of very heterogeneous datasets, the size of the condensed representation remained too large and the use of a clustering algorithm was too costly. We have introduced and used the notion of naive assignment in previous work [6], but this partitioning resulted in a high number of meaningless comparisons as a given pair of entities is compared several times. With respect to our previous work, this paper has the following enhancements: (i) a new formalization of the concepts, (ii) a complete rewriting of the algorithms and descriptions, (iii) a novel distribution principle leading to significant improvement of performances, (iv) an extensive experimental study.

Our clustering algorithm is inspired by DBSCAN, which is well suited to the requirements of RDF datasets. This is mainly because it produces clusters of arbitrary shape, which is important in our context where entities of the same type can be described by heterogeneous property sets. Furthermore, it does not require as an input the number of resulting clusters, and it detects noise points which are not important enough to form a class. However, the main weakness of DBSCAN is its computational complexity which is $\mathcal{O}(n^2)$, where n is the number of data points.

Many works have proposed approaches to scale-up the DBSCAN algorithm by parallelizing its execution. Some of these algorithms are based on a random split of the data. In PDSDBSCAN [22], the data is partitioned randomly, and the clustering is applied in each partition in parallel by comparing the entities in one partition with the whole dataset. S-DBSCAN [20] merges the clusters that have close centers. The approach proposed in [25] merges the clusters that intersect with each other based on the centers and the radius of the clusters. In [13], after partitioning the data and calculating the local clusters, a range is defined for each partition, and the points outside this range are considered as seeds to merge the local clusters. Algorithms based on a random split of the data achieve a fast clustering, but at the cost of a lower accuracy; they produce a schema of a lower quality compared to other existing approaches. The ϵ-*neighborhood* of the entities is computed in random sub-sets, neighbors in different partitions are therefore not discovered. In addition, the merging relies on features such as the cluster's center and does not ensure that the result is the same as the one of the DBSCAN algorithm.

Some works propose algorithms such as MR-DBSCAN [14] and RDD-DBSCAN [23] which partition the data using Binary Space Partitioning (BSP) [11], duplicate the frontiers of each partition into the neighboring partitions and generate the clusters. The clusters are finally merged if they share some entities. However, approaches using Binary Space Partitioning lose their efficiency when applied to data with high dimensionality such as RDF datasets.

RP-DBSCAN [26] combines different techniques, as it consists in randomly partitioning cells of data, then creating a graph using BSP to accelerate the neighbors search in each partition. Finally, it merges the clusters found in each partition to provide the final clusters. As it uses a cell-based grid structure, this algorithm can not be applied on RDF datasets because it is impossible to represent an RDF dataset in such n-dimensional space. Moreover, the quality of the resulting clusters depends on a given parameter ρ and does not always ensure that the clustering is the same as the one of DBSCAN.

Finally, some graph based approaches have been proposed such as NG-DBSCAN [19], which comprises two steps: first, it computes the ϵ-graph by comparing each point with k randomly selected points and adding an edge between the closest ones. Second, it considers the edges having the highest number of neighbors as the cluster's root and all the elements connected to this root are assigned to the same cluster. However, unlike our approach, NG-DBSCAN provides a probabilistic result which is different from the one provided by DBSCAN; this reduces the quality of the resulting schema. In addition, building the neighbor graph for large datasets is a costly operation.

10   Conclusion

In this paper, we have proposed an approach that automatically extracts the underlying schema of a large RDF dataset. It relies on a novel distributed algorithm for density-based clustering which groups the similar entities into clusters and produces the same result as the DBSCAN algorithm. The resulting clusters represent the classes of the schema.

We have implemented our algorithm using Spark, a big data technology offering a fast distributed execution of the algorithm and allowing to cluster massive datasets containing millions of entities. We have shown through detailed experiments that our algorithm provides a schema of good quality, and scales up to very large datasets, outperforming existing similar clustering algorithms.

We have used both synthetically generated datasets and real datasets extracted from DBpedia.

The schema discovery approach proposed in this paper has been designed for RDF data; however, it can be adapted and applied to data sources described using other formats such as Json or XML, where the entities are irregular and do not have a defined structure.

In our future works, we will enrich the generated schema by extracting links between the classes and constraints on the properties. We will also improve our approach by automatically detecting the most appropriate values of the parameters, such as the *capacity* parameter according to the configuration of computing nodes. Schema evolution is also an important issue to be tackled in our future works; once the schema is generated, appropriate algorithms are required to keep the schema consistent with the dataset over time, as data is added or deleted.

References

1. Abiteboul, S., et al.: Research directions for principles of data management (Dagstuhl perspectives workshop 16151). Dagstuhl Manifestos **7**(1), 1–29 (2018)
2. Alcalde, C., Burusco, A.: Study of the relevance of objects and attributes of *L*-fuzzy contexts using overlap indexes. In: Medina, J., et al. (eds.) IPMU 2018. CCIS, vol. 853, pp. 537–548. Springer, Cham (2018). https://doi.org/10.1007/978-3-319-91473-2_46
3. Auer, S., Bizer, C., Kobilarov, G., Lehmann, J., Cyganiak, R., Ives, Z.: DBpedia: a nucleus for a web of open data. In: Aberer, K., et al. (eds.) ASWC/ISWC -2007. LNCS, vol. 4825, pp. 722–735. Springer, Heidelberg (2007). https://doi.org/10.1007/978-3-540-76298-0_52
4. Baazizi, M.A., Lahmar, H.B., Colazzo, D., Ghelli, G., Sartiani, C.: Schema inference for massive JSON datasets. In: Proceeding of the 20th International Conference on Extending Database Technology (EDBT), pp. 222–233 (2017)
5. Baazizi, M.-A., Colazzo, D., Ghelli, G., Sartiani, C.: Parametric schema inference for massive JSON datasets. VLDB J. **28**(4), 497–521 (2019). https://doi.org/10.1007/s00778-018-0532-7
6. Bouhamoum, R., Kedad, Z., Lopes, S.: Schema discovery in large web data sources. In: proceeding of the 1st International Conference on Big Data and Cybersecurity Intelligence (BDCSIntell) (2018)
7. Bouhamoum, R., Kellou-Menouer, K.K., Lopes, S., Kedad, Z.: Scaling up schema discovery approaches. In: Proceeding of the 34th International Conference on Data Engineering Workshops (ICDEW), pp. 84–89. IEEE (2018)
8. Campina, S., Perry, T.E., Ceccarelli, D., Delbru, R., Tummarello, G.: Introducing RDF graph summary with application to assisted SPARQL formulation. In: Proceeding of the 23rd International Workshop on Database and Expert Systems Applications (DEXA), pp. 261–266. IEEE (2012)
9. Christodoulou, K., Paton, N.W., Fernandes, A.A.A.: Structure inference for linked data sources using clustering. In: Hameurlain, A., Küng, J., Wagner, R., Bianchini, D., De Antonellis, V., De Virgilio, R. (eds.) Transactions on Large-Scale Data- and Knowledge-Centered Systems XIX. LNCS, vol. 8990, pp. 1–25. Springer, Heidelberg (2015). https://doi.org/10.1007/978-3-662-46562-2_1

10. Ester, M., Kriegel, H.P., Sander, J., Xu, X.: A density-based algorithm for discovering clusters in large spatial databases with noise. In: Proceeding of the Second International Conference on Knowledge Discovery and Data Mining (KDD), pp. 226–231. AAAI Press (1996)
11. Fuchs, H., Kedem, Z.M., Naylor, B.F.: On visible surface generation by a priori tree structures. In: Proceedings of the 7th Annual Conference on Computer Graphics and Interactive Techniques (SIGGRAPH) pp. 124–133. ACM Press (1980)
12. Gragera Aguaza, A., Suppakitpaisarn, V.: Relaxed triangle inequality ratio of the Sørensen-dice and Tversky indexes. Theoret. Comput. Sci. **718**, 37–45 (2017)
13. Han, D., Agrawal, A., Liao, W., Choudhary, A.: A novel scalable DBSCAN algorithm with spark. In: Proceeding of the 29th International Parallel and Distributed Processing Symposium Workshops (IPDPSW), pp. 1393–1402. IEEE (2016)
14. He, Y., Tan, H., Luo, W., Feng, S., Fan, J.: MR-DBSCAN: a scalable MapReduce-based DBSCAN algorithm for heavily skewed data. Front. Comput. Sci. **8**(1), 83–99 (2014). https://doi.org/10.1007/s11704-013-3158-3. Proceeding of the 27th International Parallel and Distributed Processing Symposium Workshops (IPDPS). Springer, Berlin, Heidelberg
15. IBM: IBM quest synthetic data generator. https://sourceforge.net/projects/ibmquestdatagen/ (2015). Accessed 1 Oct 2018
16. Jaccard, P.: The distribution of flora in the Alpine zone. New Phytologist **11**(2), 37–50 (1912)
17. Kellou-Menouer, K., Kedad, Z.: Schema discovery in RDF data sources. In: Johannesson, P., Lee, M.L., Liddle, S.W., Opdahl, A.L., López, Ó.P. (eds.) ER 2015. LNCS, vol. 9381, pp. 481–495. Springer, Cham (2015). https://doi.org/10.1007/978-3-319-25264-3_36
18. Kellou-Menouer, K., Kedad, Z.: A self-adaptive and incremental approach for data profiling in the semantic web. In: Hameurlain, A., Küng, J., Wagner, R. (eds.) Transactions on Large-Scale Data- and Knowledge-Centered Systems XXIX. LNCS, vol. 10120, pp. 108–133. Springer, Heidelberg (2016). https://doi.org/10.1007/978-3-662-54037-4_4
19. Lulli, A., Dell'Amico, M., Michiardi, P., Ricci, L.: NG-DBSCAN: scalable density-based clustering for arbitrary data. Proc. VLDB Endow. **10**(3), 157–168 (2016). https://doi.org/10.14778/3021924.3021932
20. Luo, G., Luo, X., Gooch, T.F.: A parallel DBSCAN algorithm based on spark. In: Proceeding of the 6th International Conference on Big Data and Cloud Computing (BDCloud), pp. 548–553. IEEE (2016)
21. Suchanek, F.M., Kasneci, G., Weikum, G.: YAGO: a core of semantic knowledge. In: Proceedings of the 16th International Conference on World Wide Web (WWW), pp. 697–706. ACM Press (2007)
22. Patwary, M.M.A., Palsetia, D., Agrawal, A., Liao, W.K., Manne, F., Choudhary, A.: A new scalable parallel DBSCAN algorithm using the disjoint-set data structure. In: Proceedings of the International Conference on High Performance Computing, Networking, Storage and Analysis (SC), pp. 1–11. IEEE (2012)
23. Patwary, M.M.A., Palsetia, D., Agrawal, A., Liao, W.K., Manne, F., Choudhary, A.: DBSCAN on resilient distributed datasets. In: Proceedings of the International Conference on High Performance Computing and Simulation (HPCS), pp. 531–540. IEEE (2015)
24. Sevilla Ruiz, D., Morales, S.F., García Molina, J.: Inferring versioned schemas from NoSQL databases and its applications. In: Johannesson, P., Lee, M.L., Liddle, S.W., Opdahl, A.L., López, Ó.P. (eds.) ER 2015. LNCS, vol. 9381, pp. 467–480. Springer, Cham (2015). https://doi.org/10.1007/978-3-319-25264-3_35

25. Savvas, I.K., Tselios, D.: Parallelizing DBSCAN algorithm using MPI. In: Proceeding of the 25th International Conference on Enabling Technologies: Infrastructure for Collaborative Enterprises (WETICE), pp. 77–82. IEEE (2016)
26. Song, H., Lee, J.G.: RP-DBSCAN: A superfast parallel DBSCAN algorithm based on random partitioning. In: Proceedings of the International Conference on Management of Data (SIGMOD), pp. 1173–1187. ACM (2018)
27. Issa, S., Paris, P.-H., Hamdi, F., Si-Said Cherfi, S.: Revealing the conceptual schemas of RDF datasets. In: Giorgini, P., Weber, B. (eds.) CAiSE 2019. LNCS, vol. 11483, pp. 312–327. Springer, Cham (2019). https://doi.org/10.1007/978-3-030-21290-2_20
28. The Apache Software Foundation: Apache Hadoop. https://hadoop.apache.org/ (2018). Accessed 20 Oct 2018
29. The Apache Software Foundation: Apache Spark. https://spark.apache.org (2018). Accessed 20 Oct 2018
30. W3C: SPARQL query language for RDF. https://www.w3.org/TR/rdf-sparql-query/ (2013). Accessed 01 Aug 2020

Load-Aware Shedding in Stream Processing Systems

Nicoló Rivetti[1], Yann Busnel[2], and Leonardo Querzoni[3]([⊠]) (ID)

[1] Rome, Italy
[2] IMT Atlantique, IRISA, Rennes, France
yann.busnel@imt-atlantique.fr
[3] DIAG, Sapienza University of Rome, Rome, Italy
querzoni@diag.uniroma1.it

Abstract. Distributed stream processing systems are today gaining momentum as a tool to perform analytics on continuous data streams. Load shedding is a technique used to handle unpredictable spikes in the input load whenever available computing resources are not adequately provisioned. In this paper, we propose Load-Aware Shedding (LAS), a novel load shedding solution that, unlike previous works, does not rely neither on a pre-defined cost model nor on any assumption on the tuple execution duration. Leveraging *sketches*, LAS efficiently estimates the execution duration of each tuple with small error bounds and uses this knowledge to proactively shed input streams at any operator to limiting queuing latencies while dropping as few tuples as possible. We provide a theoretical analysis proving that LAS is an (ε, δ)-approximation of the optimal online load shedder. Furthermore, through an extensive practical evaluation based on simulations and a prototype, we evaluate its impact on stream processing applications.

Keywords: Load-shedding · Stream processing · Data streaming · Distributed systems

1 Introduction

Distributed stream processing systems (DSPS) and Complex Event Processing (CEP) are today considered as a mainstream technology to build architectures for the real-time analysis of big data. An application running in a DSPS, or a query executed by a CEP engine, is typically modeled as a directed acyclic graph (a topology) where data operators, represented by nodes, are interconnected by streams of tuples containing data to be analyzed, the directed edges. The success

This work has been partially funded by the MIUR SCN-00064 project RoMA and by Sapienza University of Rome through the project RM11916B75A3293D.

A preliminary short version of this work appeared in the *Proceedings of the 10th ACM International Conference on Distributed and Event-based Systems*.

N. Rivetti—Independent researcher.

© Springer-Verlag GmbH Germany, part of Springer Nature 2020
A. Hameurlain and A M. Tjoa (Eds.): TLDKS XLVI, LNCS 12410, pp. 121–153, 2020.
https://doi.org/10.1007/978-3-662-62386-2_5

of such systems can be traced back to their ability to run complex applications at scale on clusters of commodity hardware or in the cloud.

Correctly provisioning computing resources for DSPS or CEP engines however is far from being a trivial task. System designers need to take into account several factors: the computational complexity of the operators, the overhead induced by the framework, and the characteristics of the input streams. This latter aspect is often the most critical, as input data streams may unpredictably change over time both in rate and in content. Over-provisioning is not economically sensible, thus system designers are today moving toward approaches based on elastic scalability [11], where an underlying infrastructure can tune at runtime the available resources in response to changes in the workload. This represents a desirable solution when coupled with on-demand provisioning offered by many cloud platforms, but still may be affected by transient overloads [3], caused for example by unexpected load spikes, that could temporarily degrade performance below the desired SLA.

Bursty input load represents a problem for both DSPS and CEP engines as it may create unpredictable bottlenecks within the system that lead to an increase in queuing latencies, pushing the system in a state where it cannot deliver the expected quality of service (typically expressed in terms of tuple completion latency). *Load shedding* is generally considered a practical approach to handle bursty traffic. It consists of dropping a subset of incoming tuples as soon as a bottleneck is detected in the system. As such, load shedding is a solution that is complementary [24] and must coexist with resource shaping techniques (like elastic scaling), rather than being an alternative.

Existing load shedding solutions either randomly drop tuples when bottlenecks are detected [1] or apply a pre-defined model of the application and its input that allows them to deterministically take the best shedding decision [25]. In any case, all the existing solutions assume that incoming tuples all impose the same computational load. However, such assumption does not hold for many practical use cases; tuple execution duration, in fact, may depend on the tuple content itself. This is often the case whenever the receiving operator implements a logic with branches where only a subset of the incoming tuples travels through every single branch. If the computation associated with each branch generates different loads, then the execution duration will change from tuple to tuple. A tuple with a large execution duration may delay the execution of subsequent tuples in the same stream, thus increasing queuing latencies. If further tuples are enqueued with large execution durations, this may bring to the emergence of a bottleneck.

As an example, consider the reach of a tweet, *i.e.*, the number of users that may receive the re-tweets of a given tweet. This computation entails counting the number of users that have a direct and un-direct follower relationship (until a given depth) with the tweet author. Then, depending on the size of the subgraph rooted in the author node, the execution times vary. For instance, in our experiments the execution time belongs to the interval $[0.01, 70]$ ms, the most frequent execution time was 65 ms, while the average per execution time was 20 ms. In the experimental evaluation, we provide a second use-case exhibiting the same phenomena.

Based on this simple observation, we introduce Load-Aware Shedding (LAS), a novel solution for load shedding in DSPS (or CEP engines) engines. LAS gets rid of the aforementioned assumptions and provides efficient shedding aimed at matching given queuing latency targets while dropping as few tuples as possible. To reach this goal LAS leverages a smart combination of *sketch* data structures to efficiently collect at runtime information on the time needed to compute tuples. This information is used to build and maintain, at runtime, a cost model that is then exploited to take decisions on when input tuples must be shed. LAS has been designed as a flexible solution that can be applied on a per-operator basis, thus allowing developers to target specific critical stream paths in their applications. The proposed solution provides predictable per operator queuing latencies, an extremely important feature in several application scenarios where the stream processing system is expected to deliver results to users in a quasi-real-time fashion. Furthermore, LAS implements an efficient load shedding solution that perfectly fits the characteristics of settings where scarce resources are available (e.g. fog-computing). Finally, LAS can be complemented by an output quality model that allows blocking it from dropping tuples that may significantly degrade the final output quality.

The contributions provided by this paper are:

- the introduction of LAS, the first solution for load shedding in DSPS (or CEP engines) that proactively drops tuples to avoid bottlenecks without requiring a predefined cost model and without any assumption on the distribution of tuples;
- a theoretical analysis of LAS that points out how it is an (ϵ, δ)-approximation of the optimal online shedding algorithm;
- an experimental evaluation that illustrates how LAS can provide predictable queuing latencies that approximate a given threshold while dropping a small fraction of the incoming tuples.

Below, the next section states the system model we consider. Afterward, Sect. 3 details LAS whose behavior is then theoretically analyzed in Sect. 4. Section 5 reports on our experimental evaluation and Sect. 6 analyzes the related works. Finally, Sect. 7 concludes the paper.

2   System Model and Problem Definition

We consider a distributed stream processing system (DSPS) or Complex Event Processing (CEP) engine deployed on a cluster where several computing nodes exchange data through messages sent over a network (Table 1). The stream processing application (or query) executed by the DSPS (or CEP engine) can be represented by a *topology*: a directed acyclic graph interconnecting operators, represented by vertices, with data streams (DS), represented by edges. Each topology contains at least a *source*, *i.e.*, an operator connected only through outbound DSs, and a *sink*, *i.e.*, an operator connected only to inbound DSs.

Table 1. Symbols used in the text.

Symbol	Description
t	Tuple
σ	Stream of tuples
$[n]$	Universe of possible tuples
f_t	Number of occurrences of t in σ
$w(t)$	Execution duration of tuple t on operator O
$q(i)$	Queuing latency of the i-th tuple of the stream
$\mathcal{D}(j)$	Set of dropped tuples
$d(j)$	Number of dropped tuples
$\overline{Q}(j)$	Average queuing latency
τ	Average queuing latency threshold
$\hat{c}$	Estimation of the total operator execution duration
$\mathcal{F}$	`Count Min` sketch that tracks tuple frequencies
$\mathcal{W}$	`Count Min` sketch that tracks tuple cumulated execution durations
N	Window size parameter
$\mathcal{S}$	Snapshot
η	Relative error between consecutive snapshots
μ	Error threshold

Data injected by the source is encapsulated in units called tuples (or events) and each data stream is an unbounded sequence of tuples. Without loss of generality, here we assume that each tuple t is a finite set of key/value pairs that can be customized to represent complex data structures. To simplify the discussion, in the rest of this work, we deal with streams of unary tuples each representing a single non-negative integer value.

For the sake of clarity, and without loss of generality, here we restrict our model to a topology with an operator LS (*load shedder*) that decides which tuples of its outbound DS σ consumed by a downstream operator O shall be dropped. The actual positioning of LS within a real topology may be tuned, depending on where bottlenecks are expected to appear within the topology itself. Nevertheless, we assume that LS is never deployed as a source or sink in any topology. Tuples in σ are drawn from a large universe $[n] = \{1, \ldots, n\}$ and are ordered, *i.e.*, $\sigma = \langle t_1, \ldots, t_m \rangle$. Therefore $[m] = 1, \ldots, m$ is the index sequence associated with the m tuples contained in the stream σ. Both m and n are unknown. We denote with f_t the unknown frequency of tuple t, *i.e.*, the number of occurrences[1] of t in σ.

We assume that the execution duration of tuple t on operator O, denoted as $w(t)$, depends on the content of the tuple t. We simplify the model assuming

---

[1] In the data streaming literature, the frequency is the number of occurrences *not* divided by time, which differs from the classical (physics) definition [17].

that w depends on a single, fixed and known attribute value of tuple t. Cases in which this assumption does not hold, e.g. w depends on multiple attributes can be simply treated by concatenating their values and considering them as a single multiplexed attribute [5,7,15]. The probability distribution of such attribute values, as well as the function w are unknown, may differ from operator to operator and may change over time. However, we assume that subsequent changes are interleaved by a large enough time frame such that an algorithm may have a reasonable amount of time to adapt. On the other hand, the input throughput of the stream may vary, even with a large magnitude, at any time.

Let $q(i)$ be the queuing latency of the i-th tuple of the stream, $i.e.$, the time spent by the i-th tuple in the inbound buffer of operator O before being processed. Let us denote as $\mathcal{D}(j) \subseteq [j], j \leq m$, the set of dropped tuples in a stream of length m, $i.e.$, dropped tuples are thus represented in $\mathcal{D}(j)$ by their indices in $[j] \subseteq [m]$. Moreover, let $d(j) \leq j \leq m$ be the number of dropped tuples in a stream prefix of length j, $i.e.$, $d(j) = |\mathcal{D}(j)|$. Then we can define the average queuing latency as: $\overline{Q}(j) = \sum_{i \in [j] \setminus \mathcal{D}(j)} q(i)/(j - d(j))$ for all $j \in [m]$.

The goal of the load shedder is to maintain at any point in the stream the average queuing latency smaller than a given threshold τ by dropping as few tuples as possible. The quality of the shedder can be evaluated both by comparing the resulting $\overline{Q}(j)$ against τ and by measuring the number of dropped tuples $d(j)$. More formally, the load shedding problem can be defined as follows[2].

Problem 1 (Load Shedding). Given a data stream $\sigma = \langle t_1, \ldots, t_m \rangle$, find the smallest set $\mathcal{D}(j)$ such that

$$\forall j \in [m] \setminus \mathcal{D}(j), \quad \overline{Q}(j) \leq \tau.$$

3   Load Aware Shedding

This section introduces the Load-Aware Shedding algorithm by first providing an overview, then detailing some background knowledge, and finally describing the details of its functioning.

3.1   Overview

Load-Aware Shedding (LAS) is based on a simple, yet effective, idea: if we assume to know the execution duration $w(t)$ of each tuple t on the operator, then we can foresee the queuing time for each tuple of the operator input stream and then drop all tuples that will cause the queuing latency threshold τ to be violated. However, the value of $w(t)$ is generally unknown. A possible solution to this problem is to build a static cost model for tuple execution duration and then use it to proactively shed load. However, building an accurate cost model usually requires a large amount of *a priori* knowledge on the system. Furthermore, once

---

[2] This is not the only possible definition of the load shedding problem. Other variants are briefly discussed in Sect. 6.

a model has been built, it can be hard to handle changes in the system or input stream characteristics at runtime.

LAS overcomes these issues by building and maintaining at run-time a cost model for tuple execution durations. It takes shedding decision based on the estimation $\widehat{C}$ of the total execution duration of the operator: $C = \sum_{i \in [m] \setminus \mathcal{D}(m)} w(t_i)$. To do so, LAS computes an estimation $\hat{w}(t)$ of the execution duration $w(t)$ of each tuple t. Then, it computes the sum of the estimated execution durations of the tuples assigned to the operator, i.e., $\widehat{C} = \sum_{i \in [m] \setminus \mathcal{D}(m)} \hat{w}(t)$. At the arrival of the i-th tuple, subtracting from $\widehat{C}$ the (physical) time elapsed from the emission of the first tuple provides us with an estimation $\hat{q}(i)$ of the queuing latency $q(i)$ for the current tuple.

To enable this approach, LAS builds a sketch on the operator (i.e., a memory efficient data structure) that will track the execution duration of the tuples it processes. Using a sketch allows LAS to efficiently track this data independently from the amount of possibly different tuples handled by the operator. When a change in the stream or operator characteristics affects the tuples execution durations $w(t)$, i.e., the sketch content changes, the operator will forward an updated version to the load shedder, which will then be able to (again) correctly estimate the tuples execution durations. This solution does not require any a priori knowledge on the stream or system and is designed to continuously adapt to changes in the input stream or on the operator characteristics.

Shedding tuples from an incoming stream has in general a negative impact on the stream processing output quality. LAS approach is focussed on discarding tuples whose contribution to operator overload is larger, independently from their content. This approach is meaningful only under the assumption that the contribution to the stream output is the same for each input tuple. Applications, where this assumption does not hold, can be managed in LAS by building up a model for output degradation caused by shedding and then using this model to check for any candidate tuple if its contribution to the output quality is compatible with a given constraint.

3.2   Background

2-Universal Hash Functions—Our algorithm uses hash functions randomly picked from a 2-universal hash functions family. A collection $\mathcal{H}$ of hash functions $h : \{1, \ldots, n\} \rightarrow \{0, \ldots, c\}$ is said to be 2-universal if for every two different items $x, y \in [n]$, for any $h \in \mathcal{H}$, $\mathbb{P}\{h(x) = h(y)\} \leq \frac{1}{c}$, which is the probability of collision obtained if the hash function assigned truly random values to any $x \in [n]$. Carter and Wegman [4] provide an efficient method to build large families of hash functions approximating the 2-universality property.

Count Min Sketch Algorithm—Cormode and Muthukrishnan have introduced in [6] the Count Min sketch that provides, for each item t in the input stream an (ε, δ)-additive-approximation $\hat{f}_t$ of the frequency f_t. The Count Min sketch consists of a two-dimensional matrix $\mathcal{F}$ of size $r \times c$, where $r = \lceil \log \frac{1}{\delta} \rceil$ and $c = \lceil \frac{e}{\varepsilon} \rceil$. Each row is associated with a different 2-universal hash function h_i:

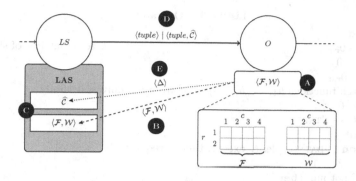

Fig. 1. Load-Aware Shedding design with $r = 2$ ($\delta = 0.25$), $c = 4$ ($\varepsilon = 0.70$).

$[n] \rightarrow [c]$. When the `Count Min` algorithm reads sample t from the input stream, it updates each row: $\forall i \in [r], \mathcal{F}[i, h_i(t)] \leftarrow \mathcal{F}[i, h_i(t)] + 1$. Thus, the cell value is the sum of the frequencies of all the items mapped to that cell. Upon request of f_t estimation, the algorithm returns the smallest cell value among the cells associated with t: $\hat{f}_t = \min_{i \in [r]} \{\mathcal{F}[i, h_i(t)]\}$.

Fed with a stream of m items, the space complexity of this algorithm is $O(\frac{1}{\varepsilon} \log \frac{1}{\delta} (\log m + \log n))$ bits, while update and query time complexities are $O(\log 1/\delta)$. The `Count Min` algorithm guarantees that the following bound holds on the estimation accuracy for each item read from the input stream: $\mathbb{P}\{|\hat{f}_t - f_t| \geq \varepsilon(m - f_t)\} \leq \delta$, while $f_t \leq \hat{f}_t$ is always true.

This algorithm can be easily generalized to provide (ε, δ)-additive-approximation of point queries on a stream of updates, *i.e.*, a stream where each item t carries a positive integer update value v_t. When the `Count Min` algorithm reads the pair $\langle t, v \rangle$ from the input stream, the update routine changes as follows: $\forall i \in [r], \mathcal{F}[i, h_i(t)] \leftarrow \mathcal{F}[i, h_i(t)] + v$.

3.3   LAS Design

The operator stores two `Count Min` sketch matrices (Fig. 1A): the first one, denoted as $\mathcal{F}$, tracks the tuple frequencies f_t; the second one, denoted as $\mathcal{W}$, tracks the tuple cumulated execution durations $W_t = w(t) \times f_t$. Both `Count Min` matrices share the same sizes, controlled by parameters ε and δ, and hash functions. The latter is the generalized version of the `Count Min` (Sect. 3.2) where the update value is the tuple execution duration when processed by the instance (*i.e.*, $v = w(t)$). The operator updates (Listing 3.1 lines 24–27) both matrices after each tuple execution.

The operator is modeled as a finite state machine (Fig. 2) with two states: START and STABILIZING. The START state lasts as long as the operator has executed N tuples, where N is a user defined window size parameter. The transition to the STABILIZING state (Fig. 2A) triggers the creation of a new snapshot $\mathcal{S}$. A snapshot is a matrix of size $r \times c$ where $\forall i \in [r], j \in [c] : \mathcal{S}[i, j] = \mathcal{W}[i, j]/\mathcal{F}[i, j]$ (Listing 3.1 lines 15–16). We say that the $\mathcal{F}$ and $\mathcal{W}$ matrices are

Listing 3.1: Operator

```
1: init do
2:      F ← 0_{r,c}                                      ▷ zero matrices of size r × c
3:      W ← 0_{r,c}
4:      S ← 0_{r,c}
5:      r hash functions h₁, ..., h_r : [n] → [c] from a 2-universal family.
6:      m ← 0
7:      state ← START
8: end init
9: function UPDATE(tuple: t, execut. time: l, request: Ĉ)
10:     m ← m + 1
11:     if Ĉ not null then
12:         Δ ← C - Ĉ
13:         send ⟨Δ⟩ to LS
14:     if state = START ∧ m  mod N = 0 then
15:                                                         ▷ Figure 2.A
16:         update S
17:         state ← STABILIZING
18:     else if state = STABILIZING ∧ m  mod N = 0 then
19:         if η ≤ μ (Eq. 1) then                          ▷ Figure 2.C
20:             send ⟨F, W⟩ to LS
21:             state ← START
22:             reset F and W to 0_{r,c}
23:         else                                           ▷ Figure 2.B
24:             update S
25:     for i = 1 to r do
26:         F[i, h_i(t)] ← F[i, h_i(t)] + 1
27:         W[i, h_i(t)] ← W[i, h_i(t)] + l
28:     end for
29: end function
```

stable when the relative error η between the previous snapshot and the current one is smaller than a parameter μ, *i.e.*,

$$\eta = \frac{\sum_{\forall i,j} |S[i,j] - \frac{W[i,j]}{F[i,j]}|}{\sum_{\forall i,j} S[i,j]} \leq \mu \tag{1}$$

is satisfied. Then, each time the operator has executed N tuples (Listing 3.1 lines 17–23), it checks whether Eq. 1 is satisfied. (i) In the negative case S is updated (Fig. 2B). (ii) In the positive case, the operator sends the F and W matrices to the load shedder (Fig. 1B), resets their content, and moves back to the START state (Fig. 2C). This approach allows to limit the amount of data sent from the operator to LS, and resembles what was proposed in [12].

There is a delay between any change in $w(t)$ and when LS receives the updated F and W matrices. This introduces a skew in the cumulated execution duration estimated by LS. To compensate this skew, we introduce a synchronization mechanism that kicks in whenever the LS receives a new pair of matrices from the operator.

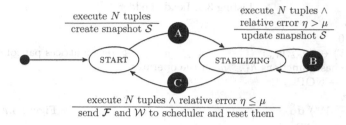

Fig. 2. Operator finite state machine.

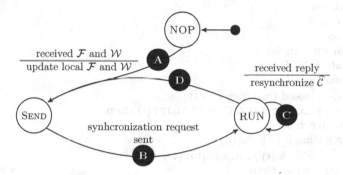

Fig. 3. Load shedder LS finite state machine.

The LS (Fig. 1C) maintains the estimated cumulated execution duration of the operator $\widehat{C}$ and a pair of initially empty matrices $\langle \mathcal{F}, \mathcal{W} \rangle$. LS is modeled as a finite state machine (Fig. 3) with three states: NOP, SEND, and RUN. The LS executes the code reported in Listing 3.2. In particular, every time a new tuple t arrives at the LS, the function SHED is executed. The LS starts in the NOP state where no action is performed (Listing 3.2 lines 15–17). Here we assume that in this initial phase, *i.e.,* when the topology has just been deployed, no load shedding is required. When LS receives the first pair $\langle \mathcal{F}, \mathcal{W} \rangle$ of matrices (Fig. 3A), it moves into the SEND state and updates its local pair of matrices (Listing 3.2 lines 7–9). While being in the SEND states, LS sends to O the current cumulated execution duration estimation $\widehat{C}$ (Fig. 1D) piggybacking it with the first tuple t that is not dropped (Listing 3.2 lines 22–24) and moves in the RUN state (Fig. 3B). This information is used to synchronize the LS with O and remove the skew between O's cumulated execution duration $\mathcal{C}$ and the estimation $\widehat{C}$ at LS. O replies to this request (Fig. 1E) with the difference $\Delta = \mathcal{C} - \widehat{C}$ (Listing 3.1 lines 11–13). When the load shedder receives the synchronization reply (Fig. 3C) it updates its estimation $\widehat{C} + \Delta$ (Listing 3.2 lines 11–13).

Listing 3.2: Load shedder

```
 1: init do
 2:     Ĉ ← 0
 3:     ⟨ℱ,𝒲⟩ ← ⟨0_{r,c}, 0_{r,c}⟩                    ▷ zero matrices pair of size r × c
 4:     Same hash functions h_1 … h_r of the operator
 5:     state ← NOP
 6: end init
 7: upon ⟨ℱ',𝒲'⟩ do                                  ▷ Figure 3.A and 3.D
 8:     state ← SEND
 9:     ⟨ℱ,𝒲⟩ ← ⟨ℱ',𝒲'⟩
10: end upon
11: upon ⟨Δ⟩ do                                       ▷ Figure 3.C
12:     Ĉ ← Ĉ + Δ
13: end upon
14: function SHED(tuple: t)
15:     if state = NOP then
16:         return false
17:     q̂ ← Ĉ− elapsed time from first tuple
18:     if CHECKLATENCY(q̂) ∧ CHECKUTILITY(t) then
19:         return true
20:     i ← arg min_{i∈[r]}{ℱ[i, h_i(t)]}
21:     Ĉ ← Ĉ + (𝒲[i, h_i(t)]/ℱ[i, h_i(t)]) × (1 + ε)
22:     if state = SEND then                           ▷ Figure 3.B
23:         piggyback Ĉ to operator on t
24:         state ← RUN
25:     return false
26: end function
27: function CHECKLATENCY(q)
28:     if (Q + q)/ℓ > τ then
29:         return true
30:     Q ← Q + q
31:     ℓ ← ℓ + 1
32:     return false
33: end function
```

In the RUN state, the load shedder computes, for each tuple t, the estimated queuing latency $\hat{q}(i)$ as the difference between the operator estimated execution duration $\hat{C}$ and the time elapsed from the emission of the first tuple (Listing 3.2 line 17). It then checks if the estimated queuing latency for t satisfies the CHECKLATENCY method (Listing 3.2 line 18).

This method encapsulates the logic for checking if a desired condition on queuing latencies is violated or not. In this paper, as stated in Sect. 2, we aim at maintaining the average queuing latency below a threshold τ. Then, CHECKLATENCY tries to add $\hat{q}$ to the current average queuing latency (Listing 3.2 lines 28). If the result is larger than τ (i), it simply returns $true$; otherwise (ii), it updates its local value for the average queuing latency and returns $false$ (Listing 3.2 lines 30–32). Note that different goals, based on the queuing latency, can be defined and encapsulated within CHECKLATENCY, $e.g.$, maintain the absolute per-tuple queuing latency below τ, or maintain the average queuing latency calculated on a sliding window below τ [21].

Function CHECKUTILITY evaluates the impact the output quality would incur by dropping t. This function encapsulates optional requirements on the maximum acceptable quality drop as defined by the semantics of the application. Considering that the quality definition is application dependent, we don't provide here a specific implementation. However, we assume that, independently of the implementation, it will return *true* if the t can be dropped with an acceptable quality loss.

If both CHECKLATENCY($\hat{q}$) and CHECKUTILITY(t) return *true* (**i**) the load shedder returns *true* as well, *i.e.*, tuple t must be dropped. Otherwise (**ii**), the operator estimated execution duration $\widehat{C}$ is updated with the estimated tuple execution duration $\hat{w}(t)$, increased by a factor $1 + \varepsilon$ to mitigate potential under-estimations[3], and the load shedder returns *false* (Listing 3.2 line 25), *i.e.*, the tuple must not be dropped. Finally, if the load shedder receives a new pair $\langle \mathcal{F}, \mathcal{W} \rangle$ of matrices (Fig. 3D), it will update its local pair of matrices and move to the SEND state (Listing 3.2 lines 7–9).

Now we will discuss the complexity of LAS.[4]

Theorem 1 (Time complexity of LAS). *For each tuple read from the input stream, the time complexity of LAS for the operator and the load shedder is* $\mathcal{O}(\log 1/\delta)$.

Theorem 2 (Space Complexity of LAS). *The space complexity of LAS for the operator and load shedder is*

$$\mathcal{O}\left(\frac{1}{\varepsilon} \log \frac{1}{\delta}(\log m + \log n)\right) \text{ bits.}$$

Theorem 3 (Communication complexity of LAS). *The communication complexity of LAS is of* $\mathcal{O}\left(\frac{m}{N}\right)$ *messages and*

$$\mathcal{O}\left(\frac{m}{N}\left(\frac{1}{\varepsilon} \log \frac{1}{\delta}(\log m + \log n) + \log m\right)\right) \text{ bits.}$$

Note that the communication cost is low with respect to the stream size since the window size N should be chosen such that $N \gg 1$ (*e.g.*, in our tests we have $N = 1024$).

4   Theoretical Analysis

This section provides an analysis of the quality of the shedding performed by LAS in two steps. First, we study the correctness and optimality of the shedding algorithm, under *full knowledge* assumption (*i.e.*, the shedding strategy is aware of the exact execution duration w_t for each tuple t). Then, in Sect. 4.2, we

---

[3] This correction factor derives from the fact that $\hat{w}(t)$ is a (ε, δ)-approximation of $w(t)$ as shown in Sect. 4.

[4] For readability reasons, proofs of these theorems are available in Appendix A.

provide a probabilistic analysis of the mechanism that LAS uses to estimate the tuple execution durations. For the sake of simplicity, in both sections, we assume CHECKUTILITY always returns *true*. The proofs of the theorem are available in Appendix A.

4.1 Correctness of LAS

We suppose that tuples cannot be preempted, that is they must be processed uninterruptedly on the available operator instance. As mentioned before, in this analysis we assume that the execution duration $w(t)$ is known for each tuple t. Finally, given our system model, we consider the problem of minimizing d, the number of dropped tuples, while guaranteeing that the average queuing latency $\overline{Q}(t)$ will be upper-bounded by τ, $\forall t \in \sigma$. The solution must work online, thus the decision of enqueueing or dropping a tuple has to be made only resorting to knowledge about tuples received so far in the stream.

Let OPT be the online algorithm that provides the optimal solution to Problem 1. We denote with $\mathcal{D}_{OPT}^{\sigma}$ (resp. d_{OPT}^{σ}) the set of dropped tuple indices (resp. the number of dropped tuples) produced by the OPT algorithm fed by stream σ (*cf.*, Sect. 2). We also denote with d_{LAS}^{σ} the number of dropped tuples produced by LAS introduced in Sect. 3.3 fed with the same stream σ.

Theorem 4 (Correctness and Optimality of LAS). *For any σ, we have* $d_{LAS}^{\sigma} = d_{OPT}^{\sigma}$ *and* $\forall t \in \sigma, \overline{Q}_{LAS}^{\sigma}(t) \leq \tau$.

This theorem establishes that LAS is optimal, given that its execution time is the same as that of the optimal OPT algorithm. Moreover, it is correct in the sense of the Definition 1 proposed in Sect. 2, namely that its average queuing latency will not exceed the predetermined threshold τ.

4.2 Execution Time Estimation

In this section, we analyze the approximation made on execution duration $w(t)$ for each tuple t when the assumption of full knowledge is removed. LAS uses two matrices, $\mathcal{F}$ and $\mathcal{W}$, to estimate the execution time $w(t)$ of each tuple submitted to the operator. By the Count Min sketch algorithm (*cf.*, Sect. 3.2) and Listing 3.1, we have that for any $t \in [n]$ and each row $i \in [r]$,

$$\mathcal{F}[i][h_i(t)](m) = f_t + \sum_{u=1, u \neq t}^{n} f_u 1_{\{h_i(u) = h_i(t)\}},$$

and

$$\mathcal{W}[i][h_i(t)](m) = f_t w_t + \sum_{u=1, u \neq t}^{n} f_u w_u 1_{\{h_i(u) = h_i(t)\}}.$$

Let us denote respectively by $w_{\min}$ and $w_{\max}$ the minimum and the maximum execution time of the items. For sake of clarity in the following equations, we denote the ratio

$$\mathcal{V}_{i,t} = \mathcal{W}[i][h_i(t)]/\mathcal{F}[i][h_i(t)].$$

We have trivially

$$w_{\min} \leq \mathcal{V}_{i,t} \leq w_{\max}.$$

We define $S = \sum_{\ell=1}^{n} w_\ell$. We then have

Theorem 5

$$\mathbb{E}\{\mathcal{V}_{i,t}\} = \frac{S - w_t}{n - 1} - \frac{k(S - nw_t)}{n(n - 1)}\left(1 - \left(1 - \frac{1}{k}\right)^n\right).$$

The proof of this theorem is available in appendix. First, it important to note that this result does not depend on m. Moreover, we easily understand that the formula proposed in this last theorem may seem rather uninformative. Thus, we propose to present a numeric application of it to take the measure of the potential use of it for an end-user.

We take for instance $k = 55$, $n = 4096$ and the distinct values of w_u equal to $1, 2, 3, \ldots, 64$, each item being present 64 times in the input stream, we get for $t = 1, \ldots, 64$, $\mathbb{E}\{\mathcal{V}_{i,t}\} \in [32.08, 32.92]$. Note also from above that we have $1 \leq \mathcal{V}_{i,t} = \mathcal{W}[i][h_i(t)]/\mathcal{F}[i][h_i(t)] \leq 64$.

From the Markov inequality, we have, for every $x > 0$,

$$\mathbb{P}\{\mathcal{V}_{i,t} \geq x\} \leq \frac{\mathbb{E}\{\mathcal{V}_{i,t}\}}{x}.$$

By taking $x = 64a$, with $a \in [0.6, 1)$, we obtain

$$\mathbb{P}\{\mathcal{V}_{i,t} \geq 64a\} \leq \frac{\mathbb{E}\{\mathcal{V}_{i,t}\}}{64a} \leq \frac{33}{64a}.$$

Recall that r denotes the number of rows of the system; we then have by the independence of the h functions,

$$\mathbb{P}\{\min_{i=1,\ldots,r}(\mathcal{V}_{i,t}) \geq 64a\}$$

$$= (\mathbb{P}\{\mathcal{V}_{i,t} \geq 64a\})^r \leq \left(\frac{33}{64a}\right)^r.$$

By taking for instance $a = 3/4$ and $r = 10$, we get

$$\mathbb{P}\{\min_{i=1,\ldots,r}(\mathcal{V}_{i,t})) \geq 48\} \leq \left(\frac{11}{16}\right)^{10} \leq 0.024.$$

5   Experimental Evaluation

In this section, we evaluate the performance obtained by using LAS to perform load shedding. We first describe the general setting used to run the tests and then discuss the results obtained through simulations (Sect. 5.2) and with a prototype of LAS integrated within Apache Storm (Sect. 5.3).

5.1   Setup

Datasets—In our tests we consider both synthetic and real datasets. Synthetic datasets are built as streams of integer values (items) representing the values of the tuple attribute driving the execution duration when processed on the operator. We consider streams of $m = 32{,}768$ tuples, each containing a value chosen among $n = 4{,}096$ distinct items. Streams have been generated using the Uniform and Zipfian distributions with different values of $\alpha \in \{0.5, 1.0, 1.5, 2.0, 2.5, 3.0\}$, denoted respectively as Zipf-0.5, Zipf-1.0, Zipf-1.5, Zipf-2.0, Zipf-2.5, and Zipf-3.0. We define w_n as the number of distinct execution duration values that the tuples can have. These w_n values are selected at a *constant* distance in the interval $[w_{min}, w_{max}]$. We ran experiments with $w_n\{1, 2, \cdots, 64\}$, however, due to space constraints, we only report results for $w_n = 64$, and with $w_{max} \in \{0.1, 0.2 \cdots, 51.2\}$ ms. Tests performed with different values for w_n did not show unexpected deviations from what is reported in this section. Unless otherwise specified, the frequency distribution is Zipf-1.0 and the stream parameters are set to $w_n = 64$, $w_{min} = 0.1$ ms and $w_{max} = 6.4$ ms; this means that the $w_n = 64$ execution durations are picked in the set $\{0.1, 0.2, \cdots, 6.4\}$ ms.

Let $\overline{W}$ be the average execution duration of the stream tuples, then the stream maximum theoretical input throughput sustainable by the setup is equal to $1/\overline{W}$. When fed with an input throughput smaller than $1/\overline{W}$ the system will be over-provisioned (*i.e.,* possible underutilization of computing resources). Conversely, an input throughput larger than $1/\overline{W}$ will result in an underprovisioned system. We refer to the ratio between the maximum theoretical input throughput and the actual input throughput as the percentage of underprovisioning that, unless otherwise stated, was set to 25%.

To generate 100 different streams, we randomize the association between the w_n execution duration values and the n distinct items: for each of the w_n execution duration values, we pick *uniformly* at random n/w_n different values in $[n]$ that will be associated to that execution duration value. This means that the 100 different streams we use in our tests do not share the same association between execution duration and item as well as the association between frequency and execution duration (thus each stream has also a different average execution duration $\overline{W}$). Each of these permutations has been run with 50 different seeds to randomize the stream ordering and the generation of the hash functions used by LAS. This means that each single experiment reports the mean outcome of 5,000 independent runs.

We considered two types of constraints defined on the queuing latency:

$\mathtt{ABS}(\tau)$: requires that the queuing latency per tuple does not exceed τ milliseconds: $\forall i \in [m] \setminus D, q(i) \leq \tau$.

$\mathtt{AVG}(\tau)$: requires that the total average queuing latency does not exceed τ milliseconds: $\forall i \in [m] \setminus D, \overline{Q}(i) \leq \tau$.

While not being a realistic requirement, the straightforwardness of the $\mathtt{ABS}(\tau)$
constraint allowed us to grasp a better insight of the mechanisms of the algorithm. However, in this section, we only show results for the $\mathtt{AVG}(6.4)$ constraint
as is it a much more sensible requirement with respect to a real setting.

The LAS operator window size parameter N, the tolerance parameter μ
and the number of rows of the $\mathcal{F}$ and $\mathcal{W}$ matrices δ were set to $N = 1024$,
$\mu = 0.05$ and $\delta = 0.1$ (*i.e.*, $r = 4$ rows) respectively. By default, the LAS precision
parameter (*i.e.*, the number of columns of the $\mathcal{F}$ and $\mathcal{W}$ matrices) was set to
$\varepsilon = 0.05$ (*i.e.*, $c = 54$ columns), however in one of the test we evaluated LAS
performance using several values: $\varepsilon \in [0.001, 1.0]$. To evaluate LAS performance
without other external factors, in all our experiments we set CHECKUTILITY to
always return *true*.

For the real data, we used a dataset containing a stream of preprocessed
tweets related to the 2014 European elections. Among other information, the
tweets are enriched with a field *mention* containing the *entities* mentioned in
the tweet. These entities can be easily classified into *politicians, media,* and
others. We consider the first 500,000 tweets, mentioning roughly $n = 35,000$
distinct entities and where the most frequent entity has an empirical probability
of occurrence equal to 0.065.

Tested Algorithms—We compare LAS performance against three other algorithms:

Base Line. The Base Linealgorithm takes as input the percentage of underprovisioning and drops at random an equivalent fraction of the tuples.

Straw-Man. The Straw-Manalgorithm uses the same shedding strategy of LAS,
however, it uses the average execution duration $\overline{W}$ as the estimated execution
duration $\hat{w}(t)$ for each tuple t.

Full Knowledge. The Full Knowledgealgorithm uses the same shedding strategy of LAS, however, it feeds it with the exact execution duration w_t for each
tuple t as they were provided by an omniscient oracle.

Evaluation Metrics—The evaluation metrics we used are:

- the dropped ratio: $\alpha = d/m$.
- the ratio of tuples dropped by algorithm alg with respect to Base Line:
  $\lambda = (d^{alg} - d^{\text{Base Line}})/d^{\text{Base Line}}$. In the following, we refer to this metric as
  shedding ratio.
- the average queuing latency: $\overline{Q} = \sum_{i \in [m] \setminus \mathcal{D}} q(i)/(m-d)$.
- the average completion latency, *i.e.*, the average time it takes for a tuple
  from the moment it is injected by the source in the topology, till the moment
  operator O concludes its processing.

Whenever applicable we provide the maximum, mean, and minimum figures over the 5,000 runs.

5.2 Simulation Results

In this section, we analyze, through a simulator built ad-hoc for this study, the sensitivity of LAS while varying several characteristics of the input load. The simulator faithfully simulates the execution of LAS and the other algorithms and simulates the execution of each tuple t on O doing busy waiting for $w(t)$ milliseconds.

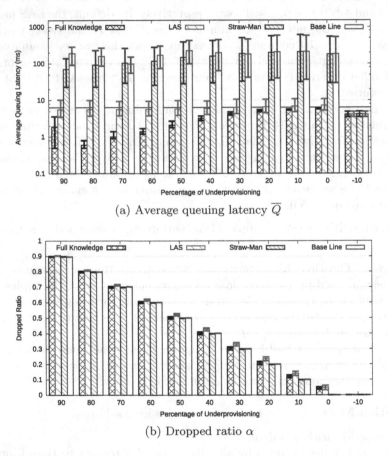

(a) Average queuing latency $\overline{Q}$

(b) Dropped ratio α

Fig. 4. LAS performance varying the amount of underprovisioning.

Input Throughput—Figure 4 shows the average queuing latency $\overline{Q}$ (top) and dropped ratio α (bottom) as a function of the percentage of under-provisioning ranging from 90% to -10% (*i.e.*, the system is 10% overprovisioned with respect

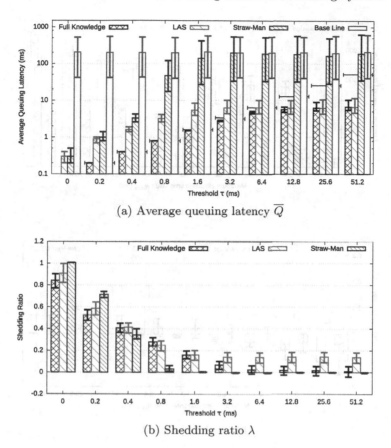

(a) Average queuing latency $\overline{Q}$

(b) Shedding ratio λ

Fig. 5. LAS performance varying the threshold τ.

to the average input throughput). As expected, in this latter case all algorithms perform at the same level as load shedding is superfluous. In all the other cases both Base Line and Straw-Man do not shed enough load and induce a huge amount of exceeding queuing latency. On the other hand, LAS average queuing latency is quite close to the required value of $\tau = 6.4$ ms, even if this threshold is violated in some of the tests. Finally, Full Knowledge always abide by the constraint and is even able to produce a much lower average queuing latency while dropping no more tuples that the competing solutions. Comparing the two plots we can see that the resulting average queuing latency is strongly linked to which tuples are dropped. In particular, Base Line and Straw-Man shed the same amount of tuples, LAS slightly more and Full Knowledge is in the middle. This result corroborates our initial claim that dropping tuples based on the load they impose allows designing more effective load shedding strategies.

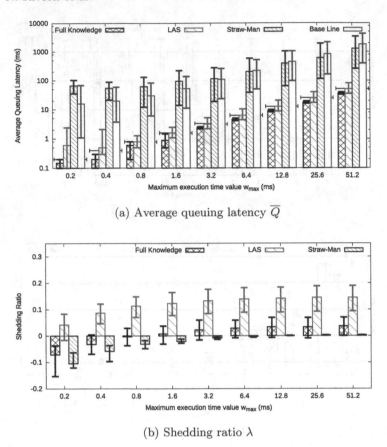

(a) Average queuing latency $\overline{Q}$

(b) Shedding ratio λ

Fig. 6. LAS performance varying the maximum execution duration value w_{max}.

Threshold τ—Figure 5 shows the average queuing latency $\overline{Q}$ (top) and shedding ratio λ (bottom) as a function of the τ threshold. Notice that with $\tau = 0$ we do not allow any queuing, while with $\tau = 6.4$ we allow at least a queuing latency equal to the maximum execution duration w_{max}. In other words, we believe that with $\tau < 6.4$ the constraint is strongly conservative, thus representing a difficult scenario for any load shedding solution. Since Base Linedoes not take into account the latency constraint τ it always drops the same amount of tuples and achieves a constant average queueing latency. For this reason, Fig. 5b reports the shedding ratio λ achieved by Full Knowledge, LAS, and Straw-Managainst Base Line. The horizontal segments in Fig. 5b represent the distinct values for τ. As the graph shows Full Knowledgealways perfectly approaches the latency threshold, but for $\tau \geq 12.8$ where it is slightly smaller. Straw-Manperforms reasonably well when the threshold is very small, but this is a consequence of the fact that it drops a large number of tuples when compared with Base Lineas can be seen by Fig. 5b. However, as τ becomes larger (*i.e.*, $\tau \geq 0.8$)

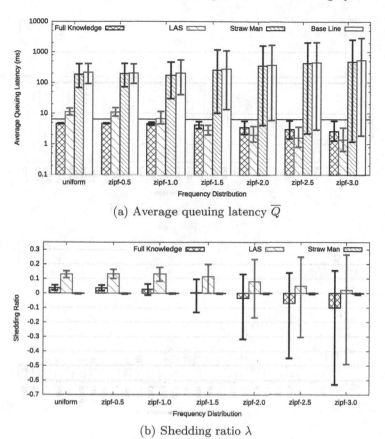

(a) Average queuing latency $\overline{Q}$

(b) Shedding ratio λ

Fig. 7. LAS performance varying the frequency probability distributions.

Straw-Manaverage queuing latency quickly grows and approaches the one from Base Lineas it starts to drop the same amount of tuples. LAS, in the same setting, performs largely better, with the average queuing latency that for large values of τ approaches the one provided by Full Knowledge. While delivering this performance LAS drops a slightly larger amount of tuples compared to Full Knowledge, to account for the approximation in calculating tuple execution durations.

Maximum Execution Duration Value w_{max}—Figure 6 shows the average queuing latency $\overline{Q}$ (top) and dropped ratio λ (bottom) as a function of the maximum execution duration value w_{max}. Notice that in this test we varied the value for τ setting it equal to w_{max}. Accordingly, Fig. 6a shows horizontal lines that mark the different thresholds τ. As the two graphs show, the behavior for LAS is rather consistent while varying w_{max}; this means that LAS can be employed in widely different settings where the load imposed by tuples in the

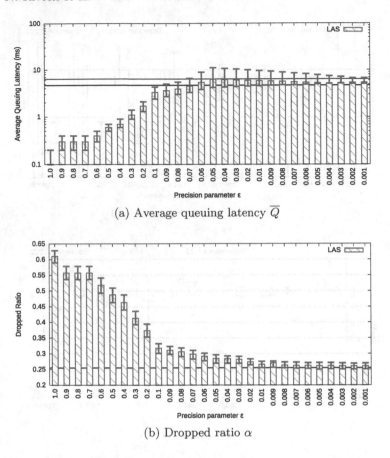

(a) Average queuing latency $\overline{Q}$

(b) Dropped ratio α

Fig. 8. LAS performance varying the precision parameter ε.

operator is not easily predictable. The price paid for this flexibility is in the shedding ratio that, as shown in Fig. 6b, is always positive.

Frequency Probability Distributions—Figure 7 shows the average queuing latency $\overline{Q}$ (top) and dropped ratio λ (bottom) as a function of the input frequency distribution. As Fig. 7a shows Straw-Manand Base Lineperform invariably bad with any distribution. The span between the best and worst performance per run increases as we move from a uniform distribution to more skewed distributions as the latter may present extreme cases where tuple latencies match their frequencies in a way that is particularly favorable or unfavorable for these two solutions. Conversely, LAS performance improves the more the frequency distribution is skewed. This result stems from the fact that the sketch data structures tracing tuple execution durations perform at their best on strongly skewed distribution, rather than on uniform ones. This result is confirmed by the shedding ratio (Fig. 7b) that decreases, on average, as α for the distribution increases.

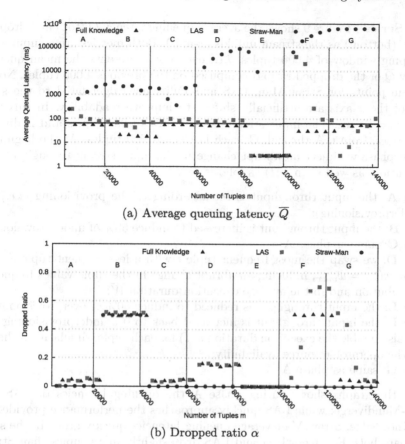

(a) Average queuing latency $\overline{Q}$

(b) Dropped ratio α

Fig. 9. Simulator time-series.

Precision Parameter ε—Figure 8 shows the average queuing latency $\overline{Q}$ (top) and dropped ratio α (bottom) as a function of the precision parameter ε. This parameter controls the trade-off between the precision and the space complexity of the sketches maintained by LAS. As a consequence, it has an impact on LAS performance. In particular, for large values of ε (left side of the graph), the sketch data structures are extremely small, thus the estimation $\hat{w}(t)$ is extremely unreliable. The corrective factor $1 + \varepsilon$ (see Listing 3.2 line 21) in this case is so large that it pushes LAS to largely overestimate the execution duration of each tuple. As a consequence LAS drops a large number of tuples while delivering average queuing latencies that are close to 0. By decreasing the value of ε (*i.e.*, $\varepsilon \leq 0.1$), sketches become larger and their estimation more reliable. In this configuration LAS performs at its best delivering average queuing latencies that are always below or equal to the threshold $\tau = 6.4$ while dropping a smaller number of tuples. The dotted lines in both graphs represent the performance of Full Knowledgeand are provided as a reference.

Time Series—Figure 9 shows the average queuing latency $\overline{Q}$ (top) and dropped ratio α (bottom) as the stream unfolds (x-axis). Both metrics are computed on a jumping window of 4.000 tuples, *i.e.,* each dot represents the mean queuing latency $\overline{Q}$ or the dropped ratio α computed on the previous 4.000 tuples. Notice that the points for Straw-Man, LAS and Full Knowledgerelated to the same value of the x-axis are artificially shifted to improve readability. In this test, we set $\tau = 64$ ms. The input stream is made of 140,000 tuples and is divided into phases, from a A through G, each lasting 20,000 tuples. At the beginning of each phase we inject an abrupt change in the input stream throughput and distribution, as well as in $w(t)$ as follows:

phase A: the input throughput is set according to the provisioning (*i.e.,* 0% underprovisioning);
phase B: the input throughput is increased to induce 50% of underprovisioning;
phase C: same as phase A;
phase D: we swap the most frequent tuple t with a less frequent tuple t' such that $w(t') = w_{max}$, inducing an abrupt change in the tuple values frequency distribution and in the average execution duration $\overline{W}$;
phase E: the input throughput is reduced to induce 50% of overprovisioning;
phase F: the input throughput is increased back to 0% underprovisioning and we also double the execution duration $w(t)$ for each tuple, simulating a change in the operator resource availability;
phase G: same as phase A.

As the graphs show, during phase A the queuing latencies of LAS and Straw-Mandiverge: while LAS quickly approaches the performance provided by Full Knowledge, Straw-Manaverage queuing latencies quickly grow. In the same timespan, both Full Knowledgeand LAS drop slightly more tuples than Straw-Man. All the three solutions correctly manage phase B: their average queuing latencies see slight changes, while, correctly, they start to drop larger amounts of tuples to compensate for the increased input throughput. The transition to phase C brings the system back in the initial configuration, while in phase D the change in the tuple frequency distribution is managed very differently by each solution: both Full Knowledgeand LAS compensate this change by starting to drop more tuples, but still maintaining the average queuing latency close to the desired threshold τ. Conversely, Straw-Mancannot handle such change, and its performance incurs a strong deterioration as it drops still the same amount of tuples. In phase E the system is strongly overprovisioned, and, as it was expected, all three solutions perform equally well as no tuple needs to be dropped. The transition to phase F is extremely abrupt as the input throughput is brought back to the equivalent of 0% of underprovisioning, but the cost to handle each tuple on the operator is doubled. At the beginning of this phase, both Straw-Manand LAS perform badly, with queuing latencies that are largely above τ. However, while the phase unfolds LAS quickly updates its data structures and converges toward the given threshold, while Straw-Mandiverges as tuples continue to be enqueued on the operator worsening the bottleneck effect. Bringing back the tuple execution durations to the initial values in phase G has little

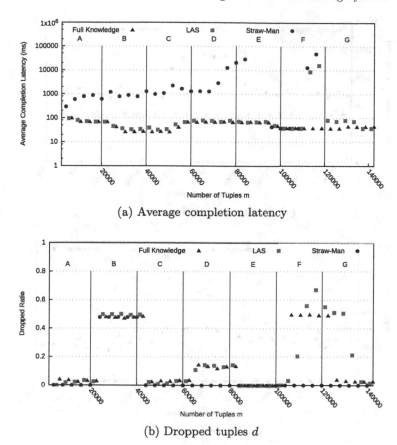

(a) Average completion latency

(b) Dropped tuples d

Fig. 10. Prototype time-series

effect on LAS, while the bottleneck created by Straw-Mancannot be recovered as it continues to drop an insufficient number of tuples.

5.3   Prototype

To evaluate the impact of LAS on real applications we implemented it as a bolt within the Apache Storm [27] framework. We have deployed our cluster on Microsoft Azure cloud service, using a Standard Tier A4 VM (4 cores and 7 GB of RAM) for each worker node, each with a single available slot.

The test topology is made of a source (*spout*) and two operators (*bolts*) LS and O. The source generates (reads) the synthetic (real) input stream and emits the tuples consumed by bolt LS. Bolt LS uses either Straw-Man, LAS or Full Knowledgeto perform the load shedding on its outbound data stream consumed by bolt O. Finally operator O implements the logic.

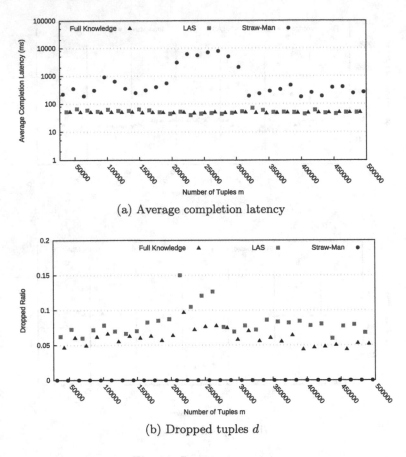

(a) Average completion latency

(b) Dropped tuples d

Fig. 11. Prototype use case

Time Series—In this test we ran the simulator using the same synthetic load used for the time series discussed in the previous section. The goal of this test is to show how our simulated tests capture the main characteristic of a real run. Notice, however, that plots in Fig. 10 report the average completion latency per tuple instead of the queuing latency. This is due to the difficulties in correctly measuring queuing latencies in Storm. Furthermore, the completion latency is, from a practical point of view, a more significant metric as it can be directly perceived on the output. From this standpoint, the results, depicted in Fig. 10, report the same qualitative behavior already discussed with Fig. 9. Two main differences are worth to be discussed: firstly, the behaviors exposed by the shedding solution in response to phase transitions in the input load are in general shifted in time (with respect to the same effects reported in Fig. 9) as a consequence of the general overhead induced by the software stack. Secondly, several data points for Straw-Manare missing in phases E and G. This is a consequence of failed tuples that start to appear as soon as the number of enqueued tuples is too large to be managed by Storm. While this may appear as a sort of "implicit" load shedding imposed by Storm, we decided not to consider these tuples in the metric calculation as they have not been dropped as a consequence of a decision taken by the Straw-Manload shedder.

Simple Application with Real Dataset—In this test we pretended to run a simple application on a real dataset: for each tweet of the twitter dataset mentioned in Sect. 5.1 we want to gather some statistics and decorate the outgoing tuples with some additional information. However, the statistics and additional information differ depending on which class the entities mentioned in each tweet belong. We assumed that this leads to a long execution duration for *media* (*e.g.*, possibly caused by access to an external DB to gather historical data), an average execution duration for *politicians* and a fast execution duration for *others* (*e.g.*, possibly because these tweets are not decorated). We modeled execution durations with 25 ms, 5 ms, and 1 ms of busy waiting respectively. Each of the 500,000 tweets may contain more than one mention, leading to $w_n = 110$ different execution duration values from $w_{min} = 1$ ms to $w_{max} = 152$ ms, among which the most frequent (36% of the stream) execution duration is 1 ms. The average execution time $\overline{W}$ is equal to 9.7 ms, the threshold τ is set to 32 ms and the under-provisioning is set to 0%.

Figure 11 reports the average completion latency (top) and dropped ratio λ (bottom) as the stream unfolds. As the plots show, LAS provides completion latencies that are extremely close to Full Knowledge, dropping a similar amount of tuples. Conversely, Straw-Mancompletion latencies are at least one order of magnitude larger. This is a consequence of the fact that in the given setting Straw-Mandoes not drop tuples, while Full Knowledgeand LAS drop on average a steady amount of tuples ranging from 5% to 10% of the stream. These results confirm the effectiveness of LAS in keeping close control on queuing latencies (and thus provide more predictable performance) at the cost of dropping a fraction of the input load.

6   Related Work

Aurora [1] is the first stream processing system where shedding has been proposed as a technique to deal with bursty input traffic. Aurora employs two different kinds of shedding, the first and better detailed being random tuple dropping at strategic places in the application topology to satisfy QoS constraints.

A large number of works proposed solutions aimed at reducing the impact of load shedding on the quality of the system output. These solutions fall under the name of *semantic* load shedding, as drop policies are linked to the significance of each tuple with respect to the computation results. Tatbul et al. first introduced in [26] the idea of semantic load shedding. Babcock et al. in [2] provided an approach tailored to aggregation queries. Tatbul et al. in [25] ported the concept of semantic load shedding in the realm of DSPS. GrubJoin [8] is a solution tailored for shedding load in multiway windowed stream joins while minimizing output degradation. Finally, Kalyvianaki et al. in [14] contextualized the problem to the realm of federated DSPS, and provided a solution for shedding fairness. Several solutions assume that the utility of an event depends on the event type and its frequency in the input event stream [26], i.e. they assume a static model for quality degradation; other works propose solutions to build and maintain at runtime a model for event utility [16,18]. All the previous works are based on the same goal, *i.e.*, to reduce the impact of load shedding on the semantics of the queries deployed in the stream processing system, while avoiding overloads. We believe that avoiding excessive degradation in the performance of the DSPS and in the semantics of the deployed query output are two orthogonal facets of the load shedding problem. In our work, we did not consider the latter and focused on the former while including in our solution the possibility to limit output quality degradation.

A different approach has been proposed in [20], with a system that builds summaries of dropped tuples to later produce approximate evaluations of queries. The idea is that such approximate results may provide users with useful information about the contribution of dropped tuples. A similar approach is adopted in StreamApprox [19] where the authors designed an online stratified reservoir sampling algorithm to produce approximate output with rigorous error bounds. A similar approach was also adopted in [28].

A classical control theory approach based on a closed control loop with feedback has been considered in [13,29,30]. In all these works the focus is on the design of the loop controller, while data is shed using a simple random selection strategy. In all these cases the goal is to reactively feed the stream processing engine system with a bounded tuple rate, without proactively considering how much load these tuples will generate.

Finally, a few works have recently appeared that address the problem of shedding load in Complex Event Processing (CEP) applications [9,10,22,23,31]. While these solution leverage techniques similar to those discussed in the previous paragraphs, they provide specific adaptations to the CEP context where intput load can be shed both in the form of events and partial pattern matches.

7 Conclusions

In this paper, we introduced Load-Aware Shedding (LAS), a novel solution for load shedding in DSPS. LAS exploits a characteristic of many stream-based applications, *i.e.*, the fact that load on operators depends both on the input rate and on the content of tuples, to smartly drop tuples and avoid the appearance of performance bottlenecks. In particular, LAS leverages sketch data structures to efficiently collect at runtime information on the operator load characteristics and then use this information to implement a load shedding policy aimed at maintaining the average queuing latencies close to a given threshold. Through a theoretical analysis, we proved that LAS is an (ϵ, δ)-approximation of the optimal algorithm. Furthermore, we extensively tested LAS both in a simulated setting, studying its sensitivity to changes of several characteristics of the input load, and with a prototype implementation integrated within the Apache Storm

DSPS. Our tests confirm that by taking into account the specific load imposed by each tuple, LAS can provide performance that closely approaches a given target, while dropping a limited number of tuples.

A   Theoretical Analysis

Data streaming algorithms strongly rely on pseudo-random functions that map elements of the stream to uniformly distributed image values to keep the essential information of the input stream, regardless of the stream elements frequency distribution.

This appendix extends with the proofs the theoretical analysis of the quality of the shedding performed by LAS in two steps provided in Sect. 4 as well as the complexities presented in Sect. 3.

First we study the correctness and optimality of the shedding algorithm, under *full knowledge* assumption (*i.e.,* the shedding strategy is aware of the exact execution duration w_t for each tuple t). Then, in Appendix A.3, we provide a probabilistic analysis of the mechanism that LAS uses to estimate the tuple execution durations.

A.1   Time, Space and Communication Complexities

In this section we provide the proofs of the time, space and communication complexities presented in Sect. 3.

Theorem 1 [Time complexity of LAS]. For each tuple read from the input stream, the time complexity of LAS for the operator and the load shedder is $\mathcal{O}(\log 1/\delta)$.

Proof. By Listing 3.1, for each tuple read from the input stream, the algorithm increments an entry per row of both the $\mathcal{F}$ and $\mathcal{W}$ matrices. Since each has $\log 1/\delta$ rows, the resulting update time complexity is $\mathcal{O}(\log 1/\delta)$. By Listing 3.2, for each submitted tuple, the scheduler has to retrieve the estimated execution duration for the submitted tuple. This operation requires to read entry per row of both the $\mathcal{F}$ and $\mathcal{W}$ matrices. Since each has $\log 1/\delta$ rows, the resulting query time complexity is $\mathcal{O}(\log 1/\delta)$.

Theorem 2 [Space Complexity of LAS]. The space complexity of LAS for the operator and load shedder is $\mathcal{O}\left(\frac{1}{\varepsilon}\log\frac{1}{\delta}(\log m + \log n)\right)$ bits.

Proof. The operator stores two matrices of size $\log(\frac{1}{\delta}) \times \frac{e}{\varepsilon}$ of counters of size $\log m$. In addition, it also stores a hash function with a domain of size n. Then the space complexity of LAS on the operator is $\mathcal{O}\left(\frac{1}{\varepsilon}\log\frac{1}{\delta}(\log m + \log n)\right)$ bits. The load shedder stores the same matrices, as well as a scalar. Then the space complexity of LAS on the load shedder is also $\mathcal{O}\left(\frac{1}{\varepsilon}\log\frac{1}{\delta}(\log m + \log n)\right)$ bits.

Theorem 3 [Communication complexity of LAS]. The communication complexity of LAS is of $\mathcal{O}\left(\frac{m}{N}\right)$ messages and $\mathcal{O}\left(\frac{m}{N}\left(\frac{1}{\varepsilon}\log\frac{1}{\delta}(\log m + \log n) + \log m\right)\right)$ bits.

Proof. After executing N tuples, the operator may send the $\mathcal{F}$ and $\mathcal{W}$ matrices to the load shedder.

This generates a communication cost of $\mathcal{O}\left(\frac{m}{N}\frac{1}{\varepsilon}\log\frac{1}{\delta}(\log m + \log n)\right)$ bits via $\mathcal{O}\left(\frac{m}{N}\right)$ messages. When the load shedder receives these matrices, the synchronization mechanism kicks in and triggers a round trip communication (half of which is piggybacked by the tuples) with the operator. The communication cost of the synchronization mechanism is $\mathcal{O}\left(\frac{m}{N}\right)$ messages and $\mathcal{O}\left(\frac{m}{N}\log m\right)$ bits.

Note that the communication cost is low with respect to the stream size since the window size N should be chosen such that $N \gg 1$ (*e.g.*, in our tests we have $N = 1024$).

A.2   Correctness of LAS

We suppose that tuples cannot be preempted, that is they must be processed uninterruptedly on the available operator instance. As mentioned before, in this analysis we assume that the execution duration $w(t)$ is known for each tuple t. Finally, given our system model, we consider the problem of minimizing d, the number of dropped tuples, while guaranteeing that the average queuing latency $\overline{Q}(t)$ will be upper-bounded by τ, $\forall t \in \sigma$. The solution must work online, thus the decision of enqueueing or dropping a tuple has to be made only resorting to knowledge about tuples received so far in the stream.

Let OPT be the online algorithm that provides the optimal solution to Problem 1. We denote with $\mathcal{D}_{OPT}^{\sigma}$ (resp. d_{OPT}^{σ}) the set of dropped tuple indices (resp. the number of dropped tuples) produced by the OPT algorithm fed by stream σ (*cf.*, Sect. 2). We also denote with d_{LAS}^{σ} the number of dropped tuples produced by LAS introduced in Sect. 3.3 fed with the same stream σ.

Theorem 4 [Correctness and Optimality of LAS]. For any σ, we have $d_{LAS}^{\sigma} = d_{OPT}^{\sigma}$ and $\forall t \in \sigma, \overline{Q}_{LAS}^{\sigma}(t) \leq \tau$.

Proof. Given a stream σ, consider the sets of indices of tuples dropped by respectively OPT and LAS, namely $\mathcal{D}_{OPT}^{\sigma}$ and $\mathcal{D}_{LAS}^{\sigma}$. Below, we prove by contradiction that $d_{LAS}^{\sigma} = d_{OPT}^{\sigma}$.

Assume that $d_{LAS}^{\sigma} > d_{OPT}^{\sigma}$. Without loss of generality, we denote $i_1, \ldots, i_{d_{LAS}^{\sigma}}$ the ordered indices in $\mathcal{D}_{LAS}^{\sigma}$, and $j_1, \ldots, j_{d_{OPT}^{\sigma}}$ the ordered indices in $\mathcal{D}_{OPT}^{\sigma}$. Let us define a as the largest natural integer such that $\forall \ell \leq a, i_\ell = j_\ell$ (*i.e.*, $i_1 = j_1, \ldots, i_a = j_a$). Thus, we have $i_{a+1} \neq j_{a+1}$.

- Assume that $i_{a+1} < j_{a+1}$. Then, according to Sect. 3.3, the i_{a+1}-th tuple of σ has been dropped by LAS as the method CHECK returned *true*. Thus, as $i_{a+1} \notin \mathcal{D}_{OPT}^{\sigma}$, the OPT run has enqueued this tuple violating the constraint τ. But this is in contradiction with the definition of OPT.

– Assume now that $i_{a+1} > j_{a+1}$. The fact that LAS does not drop the j_{a+1} tuple means that CHECK returns *false*, thus that tuple does not violate the constraint on τ. However, as OPT is optimal, it may drop some tuples for which CHECK is *false*, just because this allows it to drop an overall lower number of tuples. Therefore, if it drops this j_{a+1} tuple, it means that OPT knows the future evolution of the stream and takes a decision on this knowledge. But, by assumption, OPT is an online algorithm, and the contradiction follows.

Then, we have that $i_{a+1} = j_{a+1}$. By induction, we iterate this reasoning for all the remaining indices from $a+1$ to d^σ_{OPT}. We then obtain that $\mathcal{D}^\sigma_{OPT} \subseteq \mathcal{D}^\sigma_{LAS}$.

As by assumption $d^\sigma_{OPT} < d^\sigma_{LAS}$, we have that $\exists \ell \in \mathcal{D}^\sigma_{LAS} \setminus \mathcal{D}^\sigma_{OPT}$ such that ℓ has been dropped by LAS. This means that, with the same tuple index prefix shared by OPT and LAS, the method CHECK returned *true* when evaluated on ℓ, and OPT would violate the condition on τ by enqueuing it. That leads to a contradiction. Then, $\mathcal{D}^\sigma_{LAS} \setminus \mathcal{D}^\sigma_{OPT} = \emptyset$, and $d^\sigma_{OPT} = d^\sigma_{LAS}$.

Furthermore, by construction, LAS never enqueues a tuple that violates the condition on τ because CHECK would return *true*.

Consequently, $\forall t \in \sigma, \overline{Q}^\sigma_{LAS}(t) \leq \tau$, which concludes the proof.

A.3   Execution Time Estimation

In this section, we analyze the approximation made on execution duration $w(t)$ for each tuple t when the assumption of full knowledge is removed. LAS uses two matrices, $\mathcal{F}$ and $\mathcal{W}$, to estimate the execution time $w(t)$ of each tuple submitted to the operator. By the Count Min sketch algorithm (*cf.*, Sect. 3.2) and Listing 3.1, we have that for any $t \in [n]$ and for each row $i \in [r]$,

$$\mathcal{F}[i][h_i(t)](m) = \sum_{u=1}^{n} f_u 1_{\{h_i(u)=h_i(t)\}}$$

$$= f_t + \sum_{u=1, u \neq t}^{n} f_u 1_{\{h_i(u)=h_i(t)\}}.$$

and

$$\mathcal{W}[i][h_i(t)](m) = f_t w_t + \sum_{u=1, u \neq t}^{n} f_u w_u 1_{\{h_i(u)=h_i(t)\}},$$

Let us denote respectively by $w_{\min}$ and $w_{\max}$ the minimum and the maximum execution time of the items. We have trivially

$$w_{\min} \leq \frac{\mathcal{W}[i][h_i(t)]}{\mathcal{F}[i][h_i(t)]} \leq w_{\max}.$$

We define $S = \sum_{\ell=1}^{n} w_\ell$. We then have

Theorem 5

$$\mathbb{E}\{\mathcal{W}[i][h_i(t)]/\mathcal{F}[i][h_i(t)]\}$$
$$= \frac{S - w_t}{n - 1} - \frac{k(S - nw_t)}{n(n - 1)}\left(1 - \left(1 - \frac{1}{k}\right)^n\right).$$

It important to note that this result does not depend on m.

Proof. For any $t = 1, \ldots, n$, $\ell = 0, \ldots, n - 1$ and $A \in U_\ell(t)$, we introduce the event $B(t, \ell, A)$ defined by

$$B(t, \ell, A) = \{h_i(u) = h_i(t), \forall u \in A \text{ and }$$
$$h_i(u) \neq h_i(t), \forall u \in \{1, \ldots, n\} \setminus (A \cup \{t\})\}.$$

From the independence of the hash function h_i, we have

$$\mathbb{P}\{B(t, \ell, A)\} = \left(\frac{1}{k}\right)^\ell \left(1 - \frac{1}{k}\right)^{n-1-\ell}.$$

Let us consider the ratio

$$\mathcal{V}_{i,t} = \mathcal{W}[i][h_i(t)]/\mathcal{F}[i][h_i(t)].$$

For any $i = 0, \ldots, n$, we define

$$R_\ell(t) = \left\{\frac{f_t w_t + \sum_{u \in A} f_u w_u}{f_t + \sum_{u \in A} f_u}, A \in U_\ell(t)\right\}.$$

We have $R_0(t) = \{w_t\}$. We introduce the set $R(t)$ defined by

$$R(t) = \bigcup_{\ell=0}^{n-1} R_\ell(t).$$

Thus with probability 1,

$$\mathcal{W}[i][h_i(t)]/\mathcal{F}[i][h_i(t)] \in R(t).$$

Let $x \in R(t)$. We have

$$\mathbb{P}\{\mathcal{V}_{i,t} = x\}$$

$$= \sum_{\ell=0}^{n-1} \sum_{A \in U_\ell(t)} \mathbb{P}\{\mathcal{V}_{i,t} = x \mid B(t, \ell, A)\}\mathbb{P}\{B(t, \ell, A)\}$$

$$= \sum_{\ell=0}^{n-1} \left(\frac{1}{k}\right)^\ell \left(1 - \frac{1}{k}\right)^{n-1-\ell} \sum_{A \in U_\ell(t)} 1_{\{x = X(t,A)\}}.$$

where $X(t, A)$ is the fraction:

$$X(t, A) = \frac{f_t w_t + \sum_{u \in A} f_u w_u}{f_t + \sum_{u \in A} f_u}$$

Thus

$$\mathbb{E}\{\mathcal{V}_{i,t}\}$$

$$= \sum_{\ell=0}^{n-1} \left(\frac{1}{k}\right)^{\ell} \left(1 - \frac{1}{k}\right)^{n-1-\ell} \sum_{A \in U_\ell(t)} \sum_{x \in R(t)} x 1_{\{x = X(t,A)\}}$$

$$= \sum_{\ell=0}^{n-1} \left(\frac{1}{k}\right)^{\ell} \left(1 - \frac{1}{k}\right)^{n-1-\ell} \sum_{A \in U_\ell(t)} X(t, A).$$

Let us assume that all the f_u are equal, that is for each u, we have $f_u = m/n$. The experimental evaluation tends to show that the worst case scenario of input streams is exhibited when all the items show the same number of occurrences in the input stream. We get

$$\mathbb{P}\{\mathcal{V}_{i,t} = x\}$$

$$= \sum_{\ell=0}^{n-1} \left(\frac{1}{k}\right)^{\ell} \left(1 - \frac{1}{k}\right)^{n-1-\ell} \sum_{A \in U_\ell(t)} 1_{\{x = \frac{w_t + \sum_{u \in A} w_u}{\ell + 1}\}}$$

that concludes the proof. $\square$

References

1. Abadi, D.J., et al.: Aurora: a new model and architecture for data stream management. Int. J. Very Large Data Bases (VLDB J.) **12**(2), 120–139 (2003)
2. Babcock, B., Datar, M., Motwani, R.: Load shedding for aggregation queries over data streams. In: Proceedings of the 20th International Conference on Data Engineering (ICDE 2004), pp. 350–361. IEEE (2004)
3. Borkowski, M., Hochreiner, C., Schulte, S.: Minimizing cost by reducing scaling operations in distributed stream processing. Proc. VLDB Endow. **12**(7), 724–737 (2019)
4. Carter, J.L., Wegman, M.N.: Universal classes of hash functions. J. Comput. Syst. Sci. **18**, 143–154 (1979)
5. Cormode., G.: Sketch techniques for approximate query processing. In: Synposes for Approximate Query Processing: Samples, Histograms, Wavelets and Sketches, Foundations and Trends in Databases. NOW Publishers (2011)
6. Cormode, G., Muthukrishnan, S.: An improved data stream summary: the count-min sketch and its applications. J. Algorithms **55**, 58–75 (2005)
7. Dobra, A., Garofalakis, M., Gehrke, J., Rastogi, R.: Sketch-based multi-query processing over data streams. In: Bertino, E., et al. (eds.) EDBT 2004. LNCS, vol. 2992, pp. 551–568. Springer, Heidelberg (2004). https://doi.org/10.1007/978-3-540-24741-8_32

8. Gedik, B., Wu, K., Yu, P.S., Liu, L.: GrubJoin: an adaptive, multi-way, windowed stream join with time correlation-aware CPU load shedding. IEEE Trans. Knowl. Data Eng. **19**(10), 1363–1380 (2007)

9. He, Y., Barman, S., Naughton, J.F.: On load shedding in complex event processing. arXiv preprint arXiv:1312.4283 (2013)

10. He, Y., Barman, S., Naughton, J.F.: On load shedding in complex event processing. In: Proceedings of the 17th International Conference on Database Theory (ICDT 2014), pp. 213–224 (2014). OpenProceedings.org

11. Heinze, T., Aniello, L., Querzoni, L., Jerzak, Z.: Cloud-based data stream processing. In: Proceedings of the 8th ACM International Conference on Distributed Event-Based Systems (DEBS 2014), pp. 238–245. ACM (2014)

12. Ilarri, S., Wolfson, O., Mena, E., Illarramendi, A., Sistla, P.: A query processor for prediction-based monitoring of data streams. In: Proceedings of the 12th International Conference on Extending Database Technology: Advances in Database Technology, EDBT 2009, pp. 415–426. Association for Computing Machinery, New York (2009)

13. Kalyvianaki, E., Charalambous, T., Fiscato, M., Pietzuch, P.: Overload management in data stream processing systems with latency guarantees. In: 7th IEEE International Workshop on Feedback Computing (Feedback Computing 2012) (2012)

14. Kalyvianaki, E., Fiscato, M., Salonidis, T., Pietzuch, P.: THEMIS: fairness in federated stream processing under overload. In: Proceedings of the 2016 International Conference on Management of Data, pp. 541–553. ACM (2016)

15. Kammoun, A.: Enhancing stream processing and complex event processing systems. Ph.D. thesis, Université Jean Monnet, Saint-Etienne (2019)

16. Katsipoulakis, N.R., Labrinidis, A., Chrysanthis, P.K.: Concept-driven load shedding: reducing size and error of voluminous and variable data streams. In: 2018 IEEE International Conference on Big Data (Big Data), pp. 418–427 (2018)

17. Muthukrishnan, S.: Data Streams: Algorithms and Applications. Now Publishers Inc. (2005)

18. Olston, C., Jiang, J., Widom, J.: Adaptive filters for continuous queries over distributed data streams. In: Proceedings of the 2003 ACM SIGMOD International Conference on Management of Data, SIGMOD 2003, pp. 563–574. Association for Computing Machinery, New York (2003)

19. Quoc, D.L., Chen, R., Bhatotia, P., Fetzer, C., Hilt, V., Strufe, T.: StreamApprox: approximate computing for stream analytics. In: Proceedings of the 18th ACM/IFIP/USENIX Middleware Conference, Middleware 2017, pp. 185–197. Association for Computing Machinery, New York (2017)

20. Reiss, F., Hellerstein, J.M.: Data triage: an adaptive architecture for load shedding in TelegraphCQ. In: Proceedings of the 21st International Conference on Data Engineering (ICDE 2005), pp. 155–156. IEEE (2005)

21. Rivetti, N., Busnel, Y., Mostefaoui, A.: Efficiently summarizing data streams over sliding windows. In: Proceedings of the 14th IEEE International Symposium on Network Computing and Applications (NCA 2015), Boston, USA, Best Student Paper Award, September 2015

22. Slo, A., Bhowmik, S., Flaig, A., Rothermel, K.: pSPICE: partial match shedding for complex event processing. In: 2019 IEEE International Conference on Big Data (Big Data), pp. 372–382. IEEE (2019)

23. Slo, A., Bhowmik, S., Rothermel, K.: eSPICE: probabilistic load shedding from input event streams in complex event processing. In: Proceedings of the 20th International Middleware Conference, pp. 215–227 (2019)

24. Stanoi, I., Mihaila, G., Palpanas, T., Lang, C.: WhiteWater: distributed processing of fast streams. IEEE Trans. Knowl. Data Eng. **19**(9), 1214–1226 (2007)
25. Tatbul, N., Çetintemel, U., Zdonik, S.: Staying fit: efficient load shedding techniques for distributed stream processing. In: Proceedings of the 33rd International Conference on Very Large Data Bases, pp. 159–170. VLDB Endowment (2007)
26. Tatbul, N., Çetintemel, U., Zdonik, S., Cherniack, M., Stonebraker, M.: Load shedding in a data stream manager. In: Proceedings of the 29th International Conference on Very Large Data Bases (VLDB 2003), pp. 309–320. VLDB Endowment (2003)
27. The Apache Software Foundation. Apache Storm. http://storm.apache.org
28. Tok, W.H., Bressan, S., Lee., M.-L.: A stratified approach to progressive approximate joins. In: Proceedings of the 11th International Conference on Extending Database Technology: Advances in Database Technology, EDBT 2008, pp. 582–593. Association for Computing Machinery, New York (2008)
29. Tu, Y.-C., Liu, S., Prabhakar, S., Yao, B.: Load shedding in stream databases: a control-based approach. In: Proceedings of the 32nd International Conference on Very Large Data Bases (VLDB 2006), pp. 787–798. VLDB Endowment (2006)
30. Zhang, Y., Huang, C., Huang, C.: A novel adaptive load shedding scheme for data stream processing. In: Future Generation Communication and Networking (FGCN 2007), pp. 378–384. IEEE (2007)
31. Zhao, B., Viet Hung, N.Q., Weidlich, M.: Load shedding for complex event processing: input-based and state-based techniques. In: 2020 IEEE 36th International Conference on Data Engineering (ICDE), Dallas, TX, USA, pp. 1093–1104 (2020). https://doi.org/10.1109/ICDE48307.2020.00099

Selectivity Estimation with Attribute Value Dependencies Using Linked Bayesian Networks

Max Halford[1,2]([✉]), Philippe Saint-Pierre[1], and Franck Morvan[2]

[1] IMT Laboratory, Paul Sabatier University, Toulouse, France
maxhalford25@gmail.com
[2] IRIT Laboratory, Paul Sabatier University, Toulouse, France

Abstract. Relational query optimisers rely on cost models to choose between different query execution plans. Selectivity estimates are known to be a crucial input to the cost model. In practice, standard selectivity estimation procedures are prone to large errors. This is mostly because they rely on the so-called attribute value independence and join uniformity assumptions. Therefore, multidimensional methods have been proposed to capture dependencies between two or more attributes both within and across relations. However, these methods require a large computational cost which makes them unusable in practice. We propose a method based on Bayesian networks that is able to capture cross-relation attribute value dependencies with little overhead. Our proposal is based on the assumption that dependencies between attributes are preserved when joins are involved. Furthermore, we introduce a parameter for trading between estimation accuracy and computational cost. We validate our work by comparing it with other relevant methods on a large workload derived from the JOB and TPC-DS benchmarks. Our results show that our method is an order of magnitude more efficient than existing methods, whilst maintaining a high level of accuracy.

1 Introduction

A query optimiser is responsible for providing a good query execution plan (QEP) for incoming database queries. To achieve this, the optimiser relies on a cost model, which tells the optimiser how much a given QEP will cost. The cost model's estimates are in large part based on the selectivity estimates of each operator inside a QEP [21]. The issue is that selectivity estimation is a difficult task. In practice, huge mistakes are not exceptions but rather the norm [30]. In turn, this leads the cost model to produce cost estimates that can be wrong by several orders of magnitude [22]. The errors made by the cost model will inevitably result in using QEPs that are far from optimal in terms of memory usage and running time. Moreover, the cost model may also be used by other systems in addition to the query optimiser. For instance, service-level agreement (SLA) negotiation frameworks are based on the assumption that the cost of each

© Springer-Verlag GmbH Germany, part of Springer Nature 2020
A. Hameurlain and A Min Tjoa (Eds.): TLDKS XLVI, LNCS 12410, pp. 154–188, 2020.
https://doi.org/10.1007/978-3-662-62386-2_6

query can accurately be estimated by the cost model [49]. Cost models are also used for admission control (should the query be run or not?), query scheduling (when to run a query?), progress monitoring (how long will a query?), and system sizing (how many resources should be allocated to run the query?) [48]. Errors made by the cost model may thus have far reaching consequences. Such errors are for the most part due to the inaccuracy of the selectivity estimates.

Selectivity estimates are usually wrong because of the many simplifying assumptions that are made by the cost model. These assumptions are known to be unverified in practice. Nonetheless, they allow the use of simple methods that have a low computational complexity. For example, the *attribute value independence* (AVI) assumption, which states that attributes are independent with each other, is ubiquitous. This justifies the widespread use of one-dimensional histograms for storing the distribution of attribute values. Another assumption which is omnipresent is the *join uniformity assumption*, which states that attributes preserve their distribution when they are part of a join. Although this is a major source of error, it rationalises the use of simple formulas that surmise uniformity [43]. Producing accurate selectivity estimates whilst preserving a low computational overhead is thus still an open research problem, even though many methods from various approaches have been proposed.

The standard approach to selectivity estimation is to build a statistical synopsis of the database. The synopsis is built at downtime and is used by the cost model when the query optimiser invokes it. The synopsis is composed of statistics that summarise each relation along with its attributes. Unidimensional constructs, e.g., histograms [20], can be used to summarise single attributes, but cannot dependencies between attributes. Multidimensional methods, e.g., multivariate histograms [38], can be used to summarise the distribution of two or more attributes. However, their spatial requirement grows exponentially with the number of attributes. Moreover, they often require a non-trivial construction phase that takes an inordinate amount of time. Another approach is to use sampling, where the idea is to run a query on a sample of the database and extrapolate the selectivity [10]. Sampling works very well for single relations. The problem is that sampling is difficult to apply in the case of joins. This is because the join of sampled relations has a high probability of being empty [7]. A different approach altogether is to acknowledge that the cost model is mostly wrong, and instead learn from its mistakes so as not to reproduce them. The most successful method in this approach is DB2's so called *learning optimiser* (LEO) [44]. Such a memorising approach can thus be used in conjunction with any cost model. Although they are appealing, memorising approaches do not help in any matter when queries that have not been seen in the past are issued. What's more, they are complementary to other methods. Finally, statistical approaches based on conditional distributions seem to strike the right balance between selectivity estimation accuracy and computational requirements [17]. A conditional distribution is a way of modifying a distribution of values based on the knowledge of another value – called the conditioning value. For example, if the values of attribute B depend on those of A, then we can write $P(A, B) = P(B|A) \times P(A)$. Conditional distributions can be organised in a so-called Bayesian network [24].

Bayesian networks thus factorise a multidimensional distribution into a product of lower dimensional ones. If well chosen, these factorisations can preserve most of the information whilst consuming much less space. In [17], we proposed to use Bayesian networks to capture attribute value dependencies inside each relation of a database. The issue with Bayesian networks is their computational cost [12]. To alleviate this issue, we restricted our networks to possess tree topologies, which leads to simpler algorithms that have the benefit in linear time. The downside of using tree topologies is that our networks capture less dependencies than a general network. However, we showed in our benchmarks that our method was able to improve the overall selectivity estimation accuracy at a very reasonable cost. The downside of our work in [17] is that it completely ignores dependencies between attributes of different relations, which we address in the present work.

Bayesian networks that capture attribute value dependencies across relations have also been proposed. [16] were the first to apply them for selectivity estimation. However, they used off-the-shelf algorithms that are standard for working with Bayesian networks, but which are costly and impractical in constrained environments. [45] extended the work of [16] to address the computational cost issues. Indeed, they proposed various constraints on the network structure of each relation's Bayesian network that reduced the overall complexity. However, this still results in a global Bayesian network with a complex structure, which requires a costly inference algorithm in order to produce selectivity estimates. Although the methods from [16] and [45] enable a competitive accuracy, they both incur a costly construction phase and are too slow at producing selectivity estimates. In light of this, our goal is to capture attribute value dependencies across relations with as little an overhead as possible. With this in mind, our method thus consists in measuring the distribution of a carefully selected set of attributes before and after a join. We do so by performing a small amount of offline joins that exploits the topology of the relations. Effectively, we make use of the fact that joins mainly occur on primary/foreign key relationships, and thus warp the attribute values distribution in a predictable way. The contributions of our paper are as follows: (1) we introduce a new assumption which simultaneously softens the attribute value independence and join uniformity assumption, (2) based on our assumption, we propose an algorithm for connecting individual Bayesian networks together into what we call a *linked Bayesian network*, (3) we show how such a linked Bayesian network can be used to efficiently estimate query selectivities both within and between relations, and (4) we introduce a parameter which allows us to generalise the trade-offs induced by existing methods based on Bayesian networks.

The rest of the paper is organised as follows. Section 2 presents the related work. Section 3 introduces some notions related to Bayesian networks and summarises the work we did in [17]. Section 3.3 introduces the methodology for combining individual Bayesian networks using the topology of a database's relations. In Sect. 4, we compare our proposal with other methods on an extensive workload derived from the JOB [30] and TPC-DS [40] benchmarks. Finally, Sect. 5 concludes the paper and hints to potential follow-ups.

2   Related Work

Ever since the seminal work of Selinger et al. [43], query optimisation has largely relied on the use of cost models. Because the most important part of the cost model is the selectivity estimation module [31], a lot of good efforts have been made across the decades. [27] first proposed the use of histograms to approximate the distribution $P(x)$ of a single attribute x. Since then, a lot of work has gone into developing optimal histograms [20] that have been used ubiquitously in cutting edge cost models. Smooth versions of histograms, e.g., kernel density estimators (KDEs) [4] and wavelets [35], have also been proposed. However, these methods are based on single attributes, and as such lose in accuracy what they gain in computational efficiency. Indeed, there is no way to capture a dependency between two attributes x and y if one only has unidimensional distributions $P(x)$ and $P(y)$ available, regardless of their accuracy.

Multidimensional distributions, i.e., $P(X_1, \ldots, X_n)$, are a way to catch dependencies between attributes. Methods based on such distributions are naturally more accurate because they soften the AVI assumption. However, they require a large amount of computational resources which hinders their use in high-throughput settings. [38] first formalised the use of equi-depth multidimensional histograms and introduced an efficient construction algorithm. [42] proposed another construction algorithm based on Hilbert curves. Multidimensional KDEs have also been proposed [18], with somewhat the same complexity guarantees. In search for efficiency, [5] offered a workload-aware method where the idea is to only build histograms for attributes that are often queried together. Even though methods based on multidimensional distributions are costly, they are implemented in some database systems and are used when specified by a database administrator. However, these methods do not help whatsoever in capturing dependencies across relations, which is probably the biggest issue cost models have to deal with.

Sampling methods have also been proposed to perform selectivity estimation. The idea is to run a query on a sample of the database and extrapolate the selectivity [10]. Sampling works very well for single relations and has been adopted by some commercial database systems. However, off the shelf sampling procedures suffer from the fact that the join of sampled relations has a high probability of being empty [7]; in other words a join has to be materialised before sampling can be done. This issue can be alleviated by the use of correlated sampling [47], where a deterministic hash function is used to ensure that samples from different relations will match with each other. Another technique is to use indexes when available [32], but this is only realistic for in-memory databases. [1] also proposed heuristics for maintaining statistics of *join synopses*. Overall, sampling is an elegant selectivity estimation method, not least because it can handle complex predicates which statistical summaries cannot (e.g., regex queries). However, sampling necessarily incurs a high computational cost. Indeed even if the samples are obtained at downtime, they still have to be loaded in memory during the query optimisation phase.

Throughout the years, a lot of proposals have been made to relax the simplifying assumptions from [43]. All of these require compromises in terms of accuracy, speed, and memory usage. The general consensus is that each method shines in a particular use case, and thus combining different methods might be a good approach. [34] formalised this idea by using a maximum entropy approach. Recently, [37] proposed combining sampling and synopses. Another approach altogether is to "give up" on the cost model and instead memorise the worst mistakes it makes so as not to reproduce them in the future [44]. There have also been proposals that make use of machine learning [2,23,25,33], where a *supervised learning* algorithm is taught to predict selectivities based on features derived from a given query and the database's metadata. Recently, deep learning methods have been proposed to extract features that don't require rules written by humans. One of the most prominent papers that advocates the use of deep learning for selectivity estimation can be found in [26]. They proposed a neural network architecture, which they dubbed MSCN for *multi-set convolutional network*. Although approaches based on supervised machine learning have had great success in other domains, their performance for query selectivity estimation isn't competitive enough, yet.

Approaches that exploit attribute correlations in order to avoid storing redundant information have also been proposed. For example, [14] proposes to build a statistical interaction model that allows to determine a relevant subset of multidimensional histograms to build. In other words, they propose to build histograms when attributes are correlated, and to make the AVI assumption if not. Bayesian networks can be seen through the same lens of exploiting redundant information. Essentially, they factorise the full probability distribution into a set of conditional distributions. A conditional distribution between two attributes implies a hierarchy whereby one of the attributes determines to some extent the other. Formally, a Bayesian network is a directed acyclic graph (DAG) where each node is an attribute and each arrow implies a conditional dependency. They can be used to summarise a probability distribution by breaking it up into smaller pieces. In comparison with the supervised learning based methods mentioned in the previous paragraph, Bayesian networks are an *unsupervised learning* method. What this means is that they directly learn by looking at the data, whereas supervised methods require a workload of queries and outputs in order to learn. In [17], we proposed to use Bayesian networks for capturing attribute value dependencies inside individual relations. [16] and [45] both proposed methods for using Bayesian networks to capture attribute value dependencies between different relations. Although this leads to more accurate selectivity estimates, it requires much more computation time and is infeasible in practice. This is due to the fact that they require the use of expensive belief propagation algorithms for performing inference. Meanwhile, our method is much faster because it restricts each Bayesian network to a tree topology, which allows the use of the variable elimination algorithm. However, our method completely ignores dependencies between attributes of different relations. Our goal in this paper is to reconcile both approaches. Effectively, we want to keep the computational benefits of building and using individual Bayesian networks, but at the same time we want our method to capture some dependencies across relations.

3 Methodology

3.1 Preliminary Work

In [17], we developed a methodology for constructing Bayesian networks to model the distribution of attribute values inside each relation of a database. A Bayesian network is a probabilistic model. As such, it is used for approximating the probability distribution of a dataset. The particularity of a Bayesian network is that it uses a directed acyclic graph (DAG) in order to do so. The graph contains one node per variable, whilst each directed edge represents a conditional dependency between two variables. Therefore, the graph is a factorisation of the full joint distribution:

$$P(X_1, \ldots, X_n) \simeq \prod_{X_i \in \mathcal{X}} P(X_i \mid Parents(X_i)) \tag{1}$$

The joint distribution $P(X_1, \ldots, X_n)$ is the probability distribution over the entire set of attributes $\{X_1, \ldots, X_n\}$. Meanwhile, $Parents(X_i)$ stands for the attributes that condition the value of X_i. The distribution $P(X_i \mid Parents(X_i))$ is thus the conditional distribution of attribute X_i's value. In practice, the full distribution is inordinately large, and is unknown to us. However, the total of the sizes of the conditional distributions $P(X_i \mid Parents(X_i))$ is much smaller.

Using standard rules of probability, such as Bayes' rule and the law of total probability [24], we are able to derive from a Bayesian network any selectivity estimation problem by converting a logical query into a product of conditional probabilities. Note, however, that a Bayesian network is necessarily an approximation of the full probability distribution because it makes assumptions about the generating process of the data. Finding the right graph structure of a Bayesian network is called *structure learning* [24].

This is usually done by maximising a scoring function, which is an expensive process that scales super-exponentially with the number of variables [12]. Approximate search methods as well as integer programming solutions have been proposed [3]. In our work in [17], we proposed to use the *Chow-Liu* algorithm [11]. This algorithm has the property of finding the best tree structure where nodes are restricted to have at most one parent. The obtained tree is the best in the sense of maximum likelihood estimation. In addition to this property, the Chow-Liu algorithm only runs in $\mathcal{O}(p^2)$ time, where p is the number of variables, and is simple to implement. It works by first computing the *mutual information* between each pair of variables, which can be seen as the strength of the relation between two variables.

The next step is to find the *maximum spanning tree* (MST) using the mutual information, and thus to derive a directed graph approximating the joint probability distribution. We propose an inference process based on variable elimination algorithm [13] since inference can be done in linear time for tree. Our experiments indicated that competitors approach are much slower. Note that our inference process can further be accelerated using the Steiner tree problem [19].

In [17], we proposed a simple method which consists in building one Bayesian network per relation. On the one hand, this has the benefit of greatly reducing the computational burden in comparison with a single large Bayesian network, as is done in [16] and [45]. On the other hand, it ignores dependencies between attributes of different relations. We will now discuss how we can improve our work from [17] in order to capture some dependencies across relations.

3.2   Handling Conditional Dependencies over Joins

The task of selectivity estimation is to determine the selectivity of a query over a set of attributes X_i that to a set of relations R_j. By making the AVI assumption, this comes down to measuring individual attribute distributions and multiplying them together, as so:

$$P(X_1, \ldots, X_n) \simeq \prod_{R_j} \left(\prod_{X_i \in R_j} P(X_i) \right) \tag{2}$$

The methodology from [17] models the attribute distribution value of a database by building a tree-shaped Bayesian network for each relation. For efficiency reasons it purposefully only captures dependencies between attributes of a single relation. As such, it ignores the many dependencies that exist between attributes of different relations and that are the bane of cost models. Essentially, this method boils down to factorising the full probability distribution as so:

$$P(X_1, \ldots, X_n) \simeq \prod_{R_j} \left(\prod_{X_i \in R_j} P(X_i \mid Parent(X_i)) \right) \tag{3}$$

where $\{X_1, \ldots, X_n\}$ is the entire set of attributes over all relations and $X_i \in R_j$ are the attributes that belong to relation R_j. $Parent(X_i)$ denotes the attribute on which the distribution of X_i is conditioned – because each Bayesian network is tree-shaped, each attribute excluding the root has a single parent. Although the work from [17] ignores dependencies between attributes of different relations, it is still much more relevant that the common assumption of full attribute value independence. Our goal in this paper is to model the full probability distribution by taking into account dependencies between attributes of different relations, which can represented as:

$$P(X_1, \ldots, X_n) \simeq \prod_{X_i} P(X_i \mid Parent(X_i)) \tag{4}$$

Note that Eq. 4 captures more information than Eq. 2. Modeling the data by taking into account conditional dependencies thus guarantees that the resulting selectivity estimates are at least as accurate as when assuming independence between attributes. In [17], we made the assumption that attribute values of different relations are independent. Additionally, we assumed that each attribute value distribution remains the same when the relation it belongs to is joined with another relation. This is called the *join uniformity assumption* and is a huge source of error. Indeed, the distributions of an attribute's values before and after a join are not necessarily the same. For instance, imagine an e-commerce website where registered customers are stored in a database alongside with purchases. Each customer can make zero or more purchases whilst each purchase is made by exactly one customer. Some customers might be registered on the website but might not have made a purchase. If the customers and purchases relations are joined together, then the customers who have not made any purchase will not be included in the join. Therefore, the attributes from the customers relation will have different value distributions when joined with the purchases relation. Note however that the attribute value distributions from the purchases relation will not be modified. This stems from the fact that the join between the customers and purchases relations is a one-to-many join. We will now explain how we can use this property to capture attribute value dependencies across relations.

Let us assume we have two relations R and S that share a primary/foreign key relationship. That is, S contains a foreign key which references the primary key of R. This means that each tuple from R can be joined with zero or more tuples from S. A direct consequence is that the size of the join $R \bowtie S$ is equal to $|S|$. The join uniformity assumptions implies that the probability for a tuple r from relation R to be present in $R \bowtie S$ follows a uniform distribution. In statistical terms, that is:

$$P(r \in R \bowtie S) \sim \mathcal{U}(\frac{1}{|R|}) \tag{5}$$

Consequently, the expected number of times each tuple from R will be part of $R \bowtie S$ is $\frac{|S|}{|R|}$. Let us now denote by $P_R(A)$ the value distribution of attribute A in relation R. We will also define $P_{R \bowtie S}(A)$ as the value distribution of attribute A in join $R \bowtie S$. The join uniformity assumption thus implies that the distribution of A's values before and after R is joined with S are equal:

$$P_R(A) = P_{R \bowtie S}(A) \tag{6}$$

Furthermore, assume we have found and built the following factorised distribution over attributes A, B, and C from relation R:

$$P_R(A, B, C) \simeq P_R(A \,|\, B) \times P_R(B \,|\, C) \times P_R(C) \tag{7}$$

If we hold the join uniformity assumption to be true, then we can use the factorised distribution to estimate selectivities for queries involving A, B and C when R is joined with S without any further modification. The issue is that this is an idealised situation that has no reason to occur in practice. On the contrary, it is likely are that some tuples from R will be more or less present than others. However, we may assume that after the join the attribute value dependencies implied by our factorisation remain valid within each relation. We call this the *attribute value dependency preservation* assumption. The idea is that if attributes A, B and C are dependent in a certain way in relation R, then there is not much reason to believe that these dependencies will disappear once R is joined with S. Although this may not necessarily always occur in practice, it is still a much softer assumption than those usually made by cost models.

To illustrate, let us consider a toy database composed of the following relations: customers with attributes $\{nationality, hair, salary\}$, shops with attributes $\{name, city, size\}$, purchases with attributes $\{day\ of\ week\}$. Moreover, assume that the purchases relation has two foreign keys, one that references the primary key of customers and another which that of shops. The purchases relation can thus be seen as a fact table whilst customers and shops can be viewed as dimension tables. represented in table. In what follows we will use the shorthand C the customers relation, S for the shops relation, and P for the purchases relation.

In the customers relation, there are Swedish customers and a lot of them have blond hair. We might capture this property in a Bayesian network with the conditional distribution $P_C(hair \mid nationality)$, which indicates that hair colour is influenced by nationality. We could suppose that the fact that Swedish people have blond hair is still true once the customers relation is joined with the purchases relation. In other words, the hair colour shouldn't change the rate at which Swedish customers make purchases. However, we may rightly assume that the number of purchases will change according to the nationality of each customer. Mathematically, we are saying the following:

$$P_{C \bowtie P}(hair, nationality) = P_C(hair \mid nationality) \times P_{C \bowtie P}(nationality) \quad (8)$$

In other words, because we assume that $P_C(hair \mid nationality)$ is equal to $P_{C \bowtie P}(hair \mid nationality)$, then we know $P_{C \bowtie P}(hair, nationality)$ – i.e., we assume that their conditional distribution remains unchanged after the join. An immediate consequence is that we get to know the $P_{C \bowtie P}(hair)$ distribution for free. Indeed, by summing over the nationalities, we obtain:

$$P_{C \bowtie P}(hair) = \sum_{nationality} P_C(hair \mid nationality) \times P_{C \bowtie P}(nationality) \quad (9)$$

To demonstrate why our assumption is useful for the purpose of selectivity estimation, let us use the example data in Tables 1 and 2.

Let us say we wish to know how many purchases are made by customers who are both blond and Swedish. The straightforward way to do this is to count the number of times "Blond" and "Swedish" appear together within the join $C \bowtie P$:

$$P_{C \bowtie P}(hair = Blond, nationality = Swedish) = \frac{5}{7} \quad (10)$$

Table 1. Customers relation

Customer	Nationality	Hair
1	Swedish	Blond
2	Swedish	Blond
3	Swedish	Brown
4	American	Blond
5	American	Brown

Table 2. Purchases relation, which contains a foreign key that is related to the primary key of the customers relation

Shop	Customer
1	1
2	1
3	1
4	1
5	2
6	3
7	5

The fraction $\frac{5}{7}$ is the true amount of purchases that were made by Swedish customers with blond hair – said otherwise this is the selectivity of the query. Obtaining it requires scanning the rows resulting from the join of C with P. In practice this can be very burdensome, especially when queries involve many relations. If we assume that the join uniformity assumption holds – in other words we assume that the value distributions of *nationality* and *hair* do not change – then we can simply reuse the Bayesian network of the customers relation:

$$P_{C \bowtie P}(Blond, Swedish) \simeq P_C(Blond \mid Swedish) \times P_C(Swedish)$$

$$\simeq \frac{2}{3} \times \frac{3}{5} \tag{11}$$

$$\simeq \frac{2}{5}$$

In this case, making the join uniformity independence assumption makes us underestimate the true selectivity by 44% $(1 - \frac{2}{5} \times \frac{7}{5})$. Some of this error is due to the fact that the nationality attribute values are not distributed in the same way once C and P are joined – indeed in this toy example Swedish customers make more purchases than American ones. However, if we know the distribution of the nationality attribute values, i.e., $P_{C \bowtie P}(nationality)$, then we can enhance our estimate in the following manner:

$$P_{C \bowtie P}(Blond, Swedish) \simeq P_C(Blond \mid Swedish) \times P_{C \bowtie P}(Swedish)$$

$$\simeq \frac{2}{3} \times \frac{6}{7} \tag{12}$$

$$\simeq \frac{4}{7}$$

Now our underestimate has shrunk to 20%. The only difference with the previous equation is that we have replaced $P_C(Swedish)$ with $P_{C \bowtie P}(Swedish)$. Note that we did not have to precompute $P_{C \bowtie P}(Blond, Swedish)$. Indeed, we assumed that the dependency between nationality and hair doesn't change once

C and P are joined, which stems from our dependency preservation assumption. Note that, in our toy example, the assumption is slightly wrong because blond customers have a higher purchase rate than brown haired ones, regardless of the nationality. Regardless, our assumption is still much softer than the join uniformity and attribute value independence assumptions.

Our assumption is softer than the join uniformity assumption because it allows attribute value distributions to change after a join. Statistically speaking, instead of assuming that tuples appear in a join following a uniform distribution, we are saying that the distribution of the tuples is conditioned on a particular attribute (e.g., the nationality of the customers dictates the distribution of the customers in the join between shops and customers). We also assume that attribute value dependencies with each relation are preserved through joins (e.g., hair colour is still dependent on nationality). The insight is that in a factorised distribution, the top-most attribute is part of any query. For instance, in the distribution $P(A \mid B) \times P(B \mid C) \times P(C)$, every query involving any combination of A, B, and C will necessarily involve $P(C)$. We will now see how our newly introduced attribute value dependency preservation assumption can be used to link Bayesian networks from different relations together, and as such relax the join uniformity and attribute value independence assumptions at the same time.

3.3   Linking Bayesian Networks

As explained in the previous subsection, if a purchases relation has a foreign key that references a primary key of another relation named customers, then the distribution of purchases' attribute values will not change after joining customers and purchases. However the distribution of customers' attribute values will change if purchases' foreign key is skewed, which is always the case to some degree. If we use the method proposed by [17], then the Bayesian network built on customers would not be accurate when estimating selectivities for queries involving customers and purchases. This is because it would assume the distributions of the attribute values from customers are preserved after the join, which is a consequence of the join uniformity assumption. Moreover, because of the AVI assumption, we would not be capturing the existing dependencies between customers's attributes and purchases's attributes because their respective attributes are assumed to be independent with those of the opposite relation. On the other hand, if we join customers and purchases and build a Bayesian network on top of the join, then we will capture the cross-relation attribute value dependencies, but at too high a computational cost [16,45]. Up to now, we have only mentioned the case where there one join occurs, but the same kind of issues occur for many-way joins – including star-joins and chain-joins.

If the attribute value distributions of customers and purchases are estimated using Bayesian networks that possess a tree structure, then we only have to include the dependencies of a subset of customers's attributes with those of purchases. Specifically, we only have to include the root attribute of customers's Bayesian network into that of of purchases. Indeed, because customers's Bayesian network is a tree, then all of its nodes are necessarily linked to the root. If we know the distribution of the root attribute's values *after* customers is joined with purchases, then, by making the attribute value dependency preservation assumption earlier introduced, we automatically obtain the distribution of the rest of customers's attribute. In other words, if the distribution of an attribute's values is modified when the relation it belongs to is joined with another relation, then we assume that all the attributes that depend on it have their value distributions modified in the exact same manner. This is another way of saying that the conditional distributions remain the same.

We will show how this works on our toy database consisting of relations customers, shops, and purchases. Following the methodology from [17], we would have built one Bayesian network per relation. Each Bayesian network would necessarily have been a tree as a consequence of using the Chow-Liu algorithm [11]. Depending on the specifics of the data, we might have obtained the Bayesian networks shown in Fig. 1.

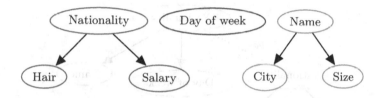

Fig. 1. Separate Bayesian networks of customers, shops, and purchases

Furthermore, let us consider the following SQL query:

```
SELECT *
FROM customers, shops, purchases
WHERE customers.id = purchases.customer_id
AND shops.id = purchases.shop_id
AND customers.nationality = 'Japanese'
AND customers.hair = 'Dark'
AND shops.name = 'Izumi'
```

If we were to estimate the amount of tuples that satisfy the above query using the Bayesian networks from Fig. 1, then we would estimate the query selectivity in the following manner:

$$P(Dark, Japanese, Izumi) = P_C(Dark \mid Japanese)$$
$$\times P_C(Japanese) \tag{13}$$
$$\times P_S(Izumi)$$

On the one hand, the conditional distribution $P_C(Dark \mid Japanese)$ captures the fact that Japanese people tend to have dark hair inside the customers relation. Graphically this is represented by the arrow that points from the "Nationality" node to the "Hair" node in Fig. 1. On the other hand, our estimate ignores the fact that shops in Japan, including "Izumi", are mostly frequented by Japanese people. The reason why is that we have one Bayesian network per relation, instead of a global network spanning all relations, and are thus not able to capture this dependency. Regardless of the missed dependency, this simple method is still more accurate than assuming total independence. Indeed the AVI assumption would neglect the dependency between *hair* and *nationality*, even though both attributes are part of the same relation. Meanwhile assuming relational independence is convenient because it only requires capturing dependencies within relations, but it discards the dependency between *nationality* and *city*. We propose to capture said dependency by adding nodes from the Bayesian networks of customers and shops to the Bayesian network of purchases. Specifically, for reasons that will become clear further on, we add the roots of the Bayesian networks of customers and shops (i.e., *nationality* and *name*) to the Bayesian network of purchases. This results in the linked Bayesian network shown in Fig. 2.

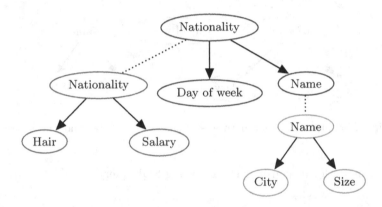

Fig. 2. Linked Bayesian network of customers, shops, and purchases

In this new configuration, we still have one Bayesian network per relation. The difference is that the Bayesian network of purchases includes the root attributes of both customers and shops's Bayesian networks. In other words, we have joined the purchases relation with the customers and shops and we have then built a Bayesian network for purchases that now includes attributes from customers and shops. A linked Bayesian network is thus a set of separate Bayesian networks where some of the attributes are duplicated in two related networks. In practice, this means that we now know the distribution of the *nationality* and *name* attribute values once the relations they belong to have been joined with purchases. Meanwhile, we also know their distributions when these relations are not joined with purchases. In other words, we store two distributions for each root attribute, one before the join and one afterwards. The distribution of a root attribute in a Bayesian network is nothing more than a one-dimensional histogram. This means that storing two distributions for each root attribute doesn't incur any significant memory burden.

The configuration shown in Fig. 2 has two immediate benefits over the one presented in Fig. 1. First of all, we are now able to determine if the percentage of Japanese in the purchases relation is different from the one in the customers relation. Indeed, we do not have to assume the distribution remains the same after the join now that we know the distribution of *nationality*'s values when customers is joined with purchases. A key observation is that we get to know something about the distribution of the *hair* attribute values when customers is joined with purchases. That is to say, because we know how the distribution of *nationality* attribute values changes after the join, then we also know something about the *hair* attribute values because both attributes are dependent within the customers relation. This stems from the fact that we assume that the conditional distribution $P(hair \mid nationality)$ is preserved after the join. Mathematically this translates to:

$$P_{C \bowtie P}(hair, nationality) = P_C(hair \mid nationality) P_{C \bowtie P}(nationality) \quad (14)$$

Although, in practice, we expect the dependency preservation assumption to not always be verified, we argue that it is a much weaker assumption than assuming total relational independence. The second benefit is that we can now take into account the fact the Japanese people typically shop in Japanese shops, even though the involved attributes belong to relations that are not directly related. This happens because the *name* attribute is now part of purchases's Bayesian network as well as that of shops. Formally the query selectivity can now be expressed as so:

$$
\begin{aligned}
P_{C \bowtie P \bowtie S}(Dark, Japanese, Izumi) &= P_C(Dark \mid Japanese) \\
&\times P_{P \bowtie S}(Izumi \mid Japanese) \quad (15) \\
&\times P_{C \bowtie P}(Japanese)
\end{aligned}
$$

Let us now consider the following SQL query where the only difference with the previous query is that are filtering by *city* instead of by *name*:

```
SELECT *
FROM customers, shops, purchases
WHERE customers.id = purchases.customer_id
AND shops.id = purchases.shop_id
AND customers.nationality = 'Japanese'
AND customers.hair = 'Dark'
AND shops.city = 'Osaka'
```

In this case, our linked Bayesian network would estimate the selectivity as so:

$$
\begin{aligned}
P(Dark, Japanese, Osaka) &= P_C(Dark \mid Japanese) \\
&\times \sum_{name} P_{P \bowtie S}(Osaka \mid name) P_P(name \mid Japanese) \\
&\times P_{C \bowtie P}(Japanese)
\end{aligned}
$$

$$(16)$$

This is a simple application of Bayesian network arithmetic [24]. The reason why there is a sum is that we have to take into account all the shops that are located in Osaka because none of them in particular has been specified in the SQL query. Note that our linked Bayesian network is still capable of estimating selectivities when only a single relation is involved. For example, we only need to use $P_P(nationality)$ when the customers relation is joined with purchases relation. If only the customers relation is involved in a query, then we can simply use $P_C(nationality)$ instead of $P_P(nationality)$. We discuss these two points in further detail in Subsect. 3.5.

Linked Bayesian networks thus combine the benefits of independent Bayesian networks, while having the benefit of softening the join uniformity assumption as well as the attribute value independence assumption. We will now discuss how one may obtain a linked Bayesian network in an efficient manner.

3.4 Building Linked Bayesian Networks

A linked Bayesian network is essentially a set of Bayesian networks. Indeed, our method consists in taking individual Bayesian networks and linking them together in order to obtain one single Bayesian network. This linking process is detailed in the next subsection. In our case, by only including the root attribute of each relation into the Bayesian network of its parent relation, we ensure that the final network necessarily has a tree topology. Performing inference on a Bayesian network with a tree topology can be done in linear time using the sum-product algorithm [29]. Building a linked Bayesian network involves building the Bayesian networks of each relation in a particular order. Indeed, in our example, we first have to build the Bayesian networks of the customers and shops relations in order to determine the roots that are to be included in the Bayesian network of the purchases relation. To build the purchases Bayesian network, we first have to join the root attributes (i.e., *nationality* and *name*) of the first two Bayesian networks (i.e., customers and shops) with the purchases relation. Naturally, performing joins incurs an added computational cost. However, we argue that joins are unavoidable if one is to capture attribute value dependencies across relations. Indeed, if joins are disallowed whatsoever, then there is basically no hope of measuring dependencies between attributes of different relations. Our methodology requires performing one left-join per primary/foreign key relationship, whilst only requiring to include one attribute per join, which is as cost-effective as possible.

The specifics of the procedure we used to build the linked Bayesian network are given in Algorithm 1. We assume the algorithm is given a set of relations. In addition, the algorithm is provided with the set of primary/foreign key relationships in the database (e.g., purchases has a foreign key that references customers' primary key and another that references shops's primary key). This set of primary/foreign key relationships can easily be extracted from any database's metadata. The idea is to go through the set of relations and check if the Bayesian networks of the dependent relations have been built. In this implementation a while loop is used to go through the relations in their topological order, from bottom to top. The Bayesian networks are built using the *BuildBN* function, which was presented in [17]. The *BuildBN* function works in three steps: 1. Build a fully-connected, undirected weighted graph, where each node is an attribute and each vertex's weight is the *mutual information* between two attributes. 2. Find the *maximum spanning tree* (MST) of the graph. 3. Orient the MST in order to obtain a tree by choosing a root.

The *BuildBN* function produces a Bayesian network with a tree topology called a *Chow-Liu tree* [11]. This tree has the property of being the tree which stores the maximum amount of information out of all the legal trees. In our algorithm, the first pass of the `while` loop will build the Bayesian networks of the relations that have no dependencies whatsoever (e.g., those who's primary key isn't referenced by any foreign key). The next pass will build the Bayesian networks of the relations that contain primary keys referenced by the foreign keys of the relations covered in the first pass. The algorithm will necessarily terminate once each relation has an associated Bayesian network; it will take as many steps as there are relations in the database.

Algorithm 1. Linked Bayesian networks construction

1: **function** BUILDLINKEDBN(*relations, relationships*)
2:     $lbn \leftarrow \{\}$
3:     $built \leftarrow \{\}$ ▷ Records which relations have been processed
4:     **while** $|lbn| < |relations|$ **do**
5:         $queue \leftarrow relations \setminus built$ ▷ Relations which don't have a BN
6:         **for each** $relation \in queue$ **do**
7:             **if** $relationships[relation] \setminus built = \varnothing$ **then**
8:                 **for each** $child \in relationships[relation]$ **do**
9:                     $relation \leftarrow relation \bowtie child.root$
10:                **end for**
11:            **end if**
12:            $lbn \leftarrow lbn \cup BuildBN(relation)$
13:            $built \leftarrow built \cup relation$
14:        **end for**
15:    **end while**
16:    **return** lbn
17: **end function**

Note that we can potentially use parallelism to speed-up the execution of Algorithm 1. Indeed, by using a priority queue and a worker pool, we can spawn processes in parallel to build the networks in the correct order. However, we consider this an implementation detail and did not take the time to implement it in our benchmark. Furthermore, this would have skewed our comparison with other methods. A linked Bayesian network doesn't require much more additional space in with respect to the method from [17]. Indeed, a linked Bayesian network is nothing more than a set of separate Bayesian networks where some of the attributes are duplicated in two related networks. Once a linked Bayesian network has been built, it can be used to produce selectivity estimates. That is, given a linked Bayesian network, we want to be able to estimate the selectivity of an arbitrary SQL query. An efficient algorithm is required to perform so-called inference when many attributes are involved, which is the topic of the following subsection.

3.5 Selectivity Estimation

The algorithm for producing selectivity estimates using linked Bayesian networks is based on the selectivity estimation algorithm proposed in [17]. The key insight is that we can fuse linked Bayesian networks into a single Bayesian network. Indeed, in our building process we have to make sure to include the root attribute of each relation's Bayesian network into its parent Bayesian's network. This allows to link each pair of adjacent Bayesian networks together via their shared attribute. In Fig. 2, these implicit links are represented with dotted lines. The purchases and customers relation have in common the *nationality* attribute, whereas the shops and purchases relations have in common the *name* attribute. The resulting "stiched" network is necessarily a tree because each individual Bayesian network is a tree and each shared attribute is located at the root of each child network.

Algorithm 2. Selectivity estimation using a linked Bayesian network

---
1: **function** INFERSELECTIVITY($lbn, query$)
2:     $relations \leftarrow ExtractRelations(query)$
3:     $relevant \leftarrow PruneLinkedBN(lbn, relations)$
4:     $linked \leftarrow LinkNetworks(relevant)$
5:     $selectivity \leftarrow ApplySumProduct(linked)$
6:     **return** $selectivity$
7: **end function**

The pseudocode for producing selectivity estimates is given in Algorithm 2. The first step of the selectivity estimation algorithm is to identify which relations are involved in a given query. Indeed each SQL query will usually involve a subset of relations, and thus we only need to use the Bayesian networks that pertain to said subset. The *PruneLinkedBN* thus takes care of removing the unnecessary Bayesian networks from the entire set of Bayesian networks. Naturally, in practice, and depending on implementation details, this may involve simply loading in memory the necessary Bayesian networks. In any case, the next step is to connect the networks into a single one. This necessitates looping over the Bayesian networks in topological order – in the same exact fashion as Algorithm 1 – and linking them along the way. Linking two Bayesian networks together simply involves replacing the attribute they have in common with the child Bayesian network. For instance, in Fig. 2, the *nationality* attribute from the purchases Bayesian network will be replaced by the customers Bayesian network. This is because we are interested in the distribution of the attributes after the join, not before. The resulting tree thus approximates the distribution of attribute values inside the (*customers* ⋈ *purchases* ⋈ *shops*) join instead of estimating selectivities inside each relation independently, as is done in textbook cost models. The result of this linking process is exemplified in Fig. 3, which shows the unrolled version of the linked Bayesian network shown in Fig. 2. Finally, once the Bayesian networks have been linked together, the sum-product algorithm [29] can be used to output the desired selectivity. In fact, this final step is exactly the same as the one described in Sect. 3.3 of [17].

Our method for estimating selectivities is very efficient. The main reason is because we only to apply the sum-product algorithm once, whereas [17] has to apply once per relation involved in the query at hand. This difference is made clear when comparing Eqs. 2 and 4. Furthermore, the sum-product algorithm is much more efficient in the case of trees than the clique tree algorithm from [45]. We confirm these insights in the benchmarks section.

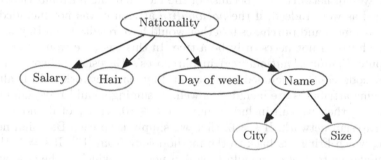

Fig. 3. Unrolled version of Fig. 2

3.6   Including More Than Just the Roots

Our model assumes that the dependencies between attribute values within a relation are preserved when a join occurs. Indeed we assume that tuples are uniformly distributed inside a join *given* each value in the root attribute. One may wonder why we have to stop at the root. Indeed, it turns out that we can include more attributes in addition to the root of each child Bayesian network when building a parent Bayesian network. For example, consider the linked Bayesian network shown in Fig. 4. In this configuration we include the *salary* attribute as well as the *nationality* attribute in the Bayesian network of the purchases relation. By doing so we obtain a new conditional distribution $P(salary|nationality)$ which tells us the dependence between *salary* and *nationality* after customers has been joined with purchases.

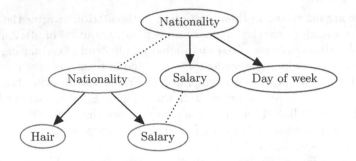

Fig. 4. Linked Bayesian network of customers and purchases

The linked Bayesian network shown in Fig. 4 is valid because we can unroll it in order to obtain a single tree, just as we did earlier on when we only included the *nationality* attribute. However, the *salary* attribute can be included in the purchases Bayesian network only because of the fact that the *nationality* attribute is included as well. Indeed, if the *nationality* attribute was not included, then linking customers and purchases together would have resulted in a Bayesian network which would not necessarily be a tree. In this case, we would not be able to compute $P(salary \mid nationality)$ in purchases's Bayesian network. In other words, a node can be included in a parent Bayesian network only if all of its conditioning attributes are included as well. Assuming a child Bayesian network has n nodes, then we can include a number $k \in \{0, \ldots, n\}$ of its nodes in the parent Bayesian network. If $k = 0$, then we simply keep each Bayesian network separate, which brings us back to the methodology from [17]. If $k = 1$, then we only include the root of each child Bayesian network, which is the case we have discussed up to now. If $k = n$, then we will include all the child's attributes in the parent BN, which is somewhat similar to the global methods presented in [16] and [45]. On the one hand, increasing k will produce larger parent Bayesian networks that capture more attribute value dependencies but also incur a higher computational cost. On the other hand, lower values of k will necessitate less computation but will assume more strongly that dependencies are preserved through joins. The k parameter is thus a practical parameter for compromising between selectivity estimation accuracy and computational requirements. Notice that different values of k can be used for each pair of relations. For instance, we might want to increase k if we notice that the cost model makes very bad estimates for a certain relation. This can be decided upon as deemed fit, be it manually or via automated DBA [46].

3.7 Summary

The method we propose attempts to generalise existing selectivity estimation methods based on Bayesian networks. Following the methodology from [17], we build one Bayesian network per relation using Chow-Liu trees. The only difference is that we include a set of attributes from the child relations into the Bayesian network associated with each parent relation. The set of included attributes depends on a chosen parameter k and the structure of each child relation's Bayesian network. Many distributions can be obtained for free because of the fact that each Bayesian network is a tree in which the root attribute conditions the rest of the attributes. This requires assuming that attribute value dependencies are preserved through joins. This assumption, although not always necessarily true, is much softer than the join uniformity as well as the attribute value independence assumptions. The resulting Bayesian networks are thus able to capture attribute value dependencies across relations, as well as inside individual relations. Although our method requires performing joins offline, we argue that joins are unavoidable if one is to capture any cross-relation dependency whatsoever. The major benefit of our method is that it only requires including a single attribute per join, and yet it brings a great deal of information for free through transitivity thanks to our newly introduced assumption. Moreover, our method can still benefit from the efficient selectivity estimation procedure presented in [17] because of the preserved tree structure. Finally, our method is able to generalise existing methods based on Bayesian networks through a single parameter which determines the amount of dependency to measure between the attributes of relations that share a primary/foreign key relationship.

4 Evaluation

4.1 Experimental Setup

We evaluate our proposal on an extensive workload derived from the JOB benchmark [30]. The JOB benchmark consists of 113 SQL queries, along with an accompanying dataset extracted from the IMDb website. The dataset consists of non-synthetic data, whereas other benchmarks such as TPC-DS [40] are based on synthetic data. The dataset is challenging because it contains skewed distributions and exhibits many correlations between attributes, both across and inside relations. The JOB benchmark is now an established and reliable standard for evaluating and comparing cost models. The dataset and the queries are publicly available[1]. In addition, we have made a Docker image available for easing future endeavours in the field[2], as well as code used in our experiments[3].

During the query optimisation phase, the cost model has to estimate the selectivity of each query execution plan (QEP) enumerated by the query optimiser. Query optimisers usually build QEPs in a bottom-up fashion [6]. Initially, the cost model will have to estimate selectivities for simple QEPs that involve a single relation. It will then be asked to estimate selectivities for larger QEPs involving multiple joins and predicates. We decided to mimic this situation by enumerating all the possible sub-queries for each of the JOB benchmark's

---

[1] JOB dataset and queries: https://github.com/gregrahn/join-order-benchmark/.

[2] Docker image: https://github.com/MaxHalford/postgres-job-docker.

[3] Method source code: https://github.com/MaxHalford/tldks-2020.

queries, as detailed in [8]. For example, if a query pertains to 4 relations, we will enumerate all the possible sub-queries involving 1, 2, 3, and all 4 relations. We also enumerate through all the combinations of filter conditions. To do so, we represented each query as a graph with each node being an attribute and each edge a join. We then simply had to retrieve all the so-called *induced subgraphs*, which are all the subgraphs that can be made up from a given graph. Each induced subgraph was then converted back to a valid SQL statement. This procedure only takes a few minutes and yields a fairly large amount of queries; indeed a total of 5,122,790 subqueries can be generated for the JOB benchmark's 113 queries. Tables 3 and 4 provide an overview of the contents of our workload.

Table 3. Query spread per number of join conditions

Joins	Amount
0	889
1–5	177,309
6–10	1,175,120
11–15	2,060,614
16–20	1,320,681
21–25	388,177

Table 4. Query spread per number of filter conditions

Filters	Amount
1	261,440
2	763,392
3	1,301,840
4	1,380,329
5	923,481
6	384,285
7	94,855
8	12,496
9	672

The general goal of our experiments is to detail the pros and cons of our method with respect to the textbook approach from [43] and some state-of-the-art methods that we were able to implement. Most industrial databases still resort to using textbook approaches, which are thus important to be compared with. Specifically our experiments solely focus on the selectivity estimation module, not on the final query execution time. We assume that improving the selectivity estimates will necessarily have a beneficial impact on the accuracy of the cost model and thus on the query execution time. Naturally, the estimation has to remain reasonable. This seems to be a view shared by many in the query optimisation community [30]. Indeed, many papers that deal with selectivity estimation, both established and new, do not measure the impact on the final query execution [9,15,41,42,45,47].

We compared our proposal with a few promising state-of-the-art methods as well as the cardinality estimation module from the PostgreSQL database system. PostgreSQL's cardinality estimation module is a fair baseline as it is a textbook implementation of the decades old ideas from [43]. We used version 10.5 of PostgreSQL and did not tinker with the default settings. Additionally, we did not bother with building indexes, as these have no consequence on the selectivity estimation module. A viable selectivity estimation method should be at least as accurate as PostgreSQL, without introducing too much of a computational cost increase. We implemented basic random sampling [39], which consists in executing a given query on a sample of each relation in order to extrapolate a selectivity estimate. Basic random sampling is simple to implement, but isn't suited for queries that involve joins because of the empty-join problem, as explained in Sect. 2. However many sampling methods that take into account the empty-join problem have been proposed. We implemented one such method, namely *correlated sampling* [47]. Correlated sampling works by hashing related primary and foreign keys and discards the tuples of linked relation where the hashes disagree. We also implemented MSCN, which is the deep learning method that is presented in [26]. Finally we implemented the Bayesian network approach from [45]. The latter method differs from ours in that it is a global approach that builds one single Bayesian networks over the entire set of relations. Although a global approach is able to capture more correlations than ours, it require more computation. We compared our method with different values for the k parameter presented in Sect. 3.6. Note that choosing $k = 0$ is equivalent to using the method from [17]. Increasing k is expected to improve the accuracy of the selectivity estimates but deteriorates the computational performance. The k parameter can thus be used to trade between accuracy and computational resources depending on the use case and the constraints of the environment.

4.2   Selectivity Estimation Accuracy

We first measured the accuracy of the selectivity estimates for each method by comparing their estimates with the true selectivity. The true selectivity can be obtained by executing the query and counting the number of tuples in the result. The appropriate metric for such a comparison is called the q-error [31, 36], and is defined as so:

$$q(y, \hat{y}) = \frac{max(y, \hat{y})}{min(y, \hat{y})} \tag{17}$$

where y is the true selectivity and $\hat{y}$ is the estimated selectivity. The q-error thus simply measures the multiplicative error between the estimate and the truth. The q-error has the property of being symmetric, and will thus be the same whether $\hat{y}$ is an underestimation or an overestimation. Moreover the q-error is scale agnostic (e.g., $\frac{8}{3} = \frac{24}{9}$), which helps in comparing errors over results with different scales.

Figure 5 shows the q-errors made by each method for all the queries of the workload derived from the JOB benchmark. The y axis represents the q-error associated with each query. Meanwhile the x axis denotes the amount of queries that have less than a given q-error. For instance, PostgreSQL managed to estimate the selectivity of two million queries with a q-error of less than 10 for each query. The curves thus give us a detailed view into the distribution of the q-errors for each method. While the curves seem to exhibit a linear trend, one

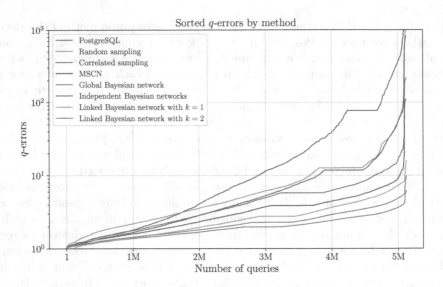

Fig. 5. Sorted q-errors for all queries by method on the JOB workload

must note that the scale of the y axis is logarithmic. The figure gives us a global idea of the accuracy of each method in comparison with the others. The mean, maximum, and meaningful quantiles of the q-errors are given in Table 5.

Table 5. q-error statistics for each method on the JOB workload

	Median	90th	95th	99th	Max	Average
PostgreSQL	7.32	77.01	185.84	707.21	10906.17	77.01
Sampling	4.79	16.45	33.17	81.34	1018.43	12.71
Correlated sampling	3.83	9.63	12.63	22.72	214.1	5.79
MSCN	2.99	6.12	7.47	12.49	110.56	3.89
Global BN	1.95	2.92	3.22	4.01	7.45	1.99
Independent BN	4.0	15.36	32.9	76.91	820.46	11.82
Linked BN $k = 1$	2.41	5.03	6.15	8.07	21.09	2.79
Linked BN $k = 2$	2.13	3.7	4.26	5.23	12.6	2.3

The overall worst method is the cost model used by PostgreSQL. This isn't a surprise, as it assumes total independence between attributes, both within and between relations. It is interesting to notice that the q-errors made by PostgreSQL's cost model can be extremely high, sometimes even reaching the tens of thousands. In this case, the query optimiser is nothing short from blind because the selectivity estimates are extremely unreliable. Although this doesn't necessarily mean that the query optimiser will not be able to find a good query execution plan, it does imply that finding a good execution plan would be down to luck [31]. One may even wonder if estimating a selectivity by picking a random number between 0 and 1 might do better. Using our method with k equal to 0 is equivalent to the methodology proposed by [17]. Indeed, if no attributes are shared by the Bayesian networks of each relation, then it is as if we considered attribute value dependencies within each relation but not between relations. As expected, the performance is similar to that of random sampling because both methods capture dependencies within a relation but not between relations. Correlated sampling performs a bit better because it is a join-aware sampling method. However, the rest of the implemented methods seems to be more precise by an order of magnitude. The deep learning method, MSCN, outperforms correlated sampling, but it isn't as performant as the Bayesian networks. However, it can probably reach a better level of performance by tuning some of the many parameters that it exposes. Meanwhile, the method we proposed with $k = 1$ means that we include the root attribute of each child relation within the Bayesian network of each parent relation. This brings to the table the benefits detailed in Sect. 3. If $k = 2$, then an additional attribute from each child relation is included with the Bayesian network of each parent relation. We can see on Fig. 6 that the global accuracy increases with k, which is what one would expect. The most accurate method overall is the global Bayesian network presented in [45]. However, our method with $k = 2$ is not far off. This makes the case that our attribute value dependency preservation assumption is a realistic one.

We have also benchmarked the methods on the TPC-DS benchmark. In contrast to the IMDb dataset used in the JOB benchmark, the TPC-DS dataset is synthetic. By nature, it contains less attribute dependencies than would be expected in a realistic use case. The TPC-DS dataset is therefore less realistic than the JOB benchmark. To produce a workload as we did for the JOB benchmark, we have taken the 30 first queries that are provided with the TPC-DS dataset and have generated all possible sub-queries. This led to a total 1,414,593 queries. The amount of joins went from 2 to 15. The overall results are shown in Table 6. As expected, the q-errors for the TPC-DS benchmark are better across the board because the dataset exhibits less correlations between attributes. Nonetheless, the rankings between the methods remains somewhat the same. Our method very slightly outperforms the global Bayesian network, but we believe that this is just an implementation artifact. In any case, our method is much more accurate than any method that assumes independence between attributes of different relations. Even so, a viable selectivity estimation method also has to be able to produce estimates in a very short amount of time, which is a point we will now discuss.

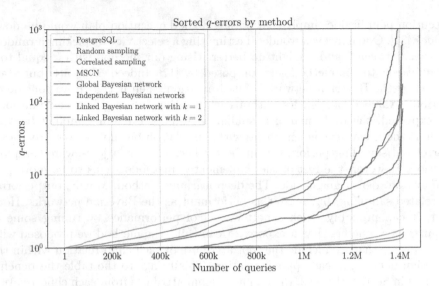

Fig. 6. Sorted q-errors for all queries by method on the TPC-DS workload

Table 6. q-error statistics for each method on the TPC-DS workload

	Median	90th	95th	99th	Max	Average
PostgreSQL	1.23	43.4	138.05	1025.49	82898.28	91.46
Sampling	3.69	13.39	26.34	66.83	669.58	9.87
Correlated sampling	2.89	8.24	10.63	19.32	170.63	4.51
MSCN	1.82	4.23	5.32	9.24	78.01	2.54
Global BN	1.03	1.17	1.23	1.33	1.49	1.06
Independent BN	2.5	13.59	33.87	89.22	597.0	9.12
Linked BN $k = 1$	1.04	1.38	1.54	1.74	1.94	1.11
Linked BN $k = 2$	1.05	1.26	1.35	1.47	1.6	1.08

4.3   Inference Time

Naturally, we next sought to measure how fast each method was at producing selectivity estimates. In a high throughput environment, the query optimiser isn't allowed to spend much time searching for an efficient QEP. In addition to using the cost model, the query optimiser also has to enumerate potential query execution plans and pick one of them [6]. Thus, only a fraction of the short amount of time allocated to the query optimiser can actually be consumed by the cost model. This means that any viable selectivity estimation has to be extremely efficient, and is probably the main reason why current cost models are kept simple. We call the amount of time necessary to produce a selectivity estimate the *inference time*. During our experiments we recorded the inference time for each query and for each model. The results shown in Table 7 show the average inference time for each method, aggregated by the number of joins present in each query.

Table 7. Average inference time in milliseconds for each method with respect to the number of joins on the JOB workload

	No joins	1 join	2 to 5 joins	6 joins or more
PostgreSQL	2.3 ± 1.1	2.6 ± 1.4	3.6 ± 1.3	8.4 ± 3.1
Sampling	19.6 ± 5.4	36.2 ± 6.8	120.2 ± 5.9	268.4 ± 8.7
Correlated sampling	20.4 ± 4.9	155.7 ± 3.2	280.6 ± 7.1	493.4 ± 9.9
MSCN	135.9 ± 12.1	312.2 ± 24.4	343.3 ± 27.4	387.6 ± 29.2
Global BN	84.3 ± 2.1	116.1 ± 2.9	145.8 ± 4.4	236.1 ± 3.8
Independent BN	8.3 ± 1.8	10.9 ± 1.3	12.6 ± 2.4	12.1 ± 3.2
Linked BN $k = 1$	9.5 ± 1.9	12.8 ± 1.6	14.1 ± 2.8	15.2 ± 3.4
Linked BN $k = 2$	10.1 ± 1.4	12.9 ± 1.5	14.3 ± 2.1	16.4 ± 2.9

It is important to mention that the inference time measured for PostgreSQL is simply the time it takes the database to execute the ANALYZE statement for each query. This thus includes the optimisation time, on top of the time spent at estimating selectivities. Even though they are already by far the best, the numbers displayed in our benchmark for PostgreSQL are pessimistic and are expected to be much lower in practice. It is also worth mentioning that we implemented the rest of the methods in Python, which is an interpreted language and thus slower than compiled languages such as C, in which PostgreSQL is written. If these methods were implemented in optimised C they would naturally be much faster. However, what matters here is the relative differences between each method, not the absolute ones.

We can clearly see from the results in Table 7 that the global Bayesian network loses in speed what it gains in accuracy. This is because it uses a complex inference method called the *clique-tree algorithm*, which is the standard approach for Bayesian networks with arbitrary topologies. Although it is the most accurate method, it is much slower than our method, regardless of the k parameter we use. What's more, the inference time of our method doesn't increase dramatically when the number of joins increases. This is due to the fact that we use a lightweight inference algorithm called *variable elimination* [13] also used by [17]. The inference algorithm scales well because we are able to merge the Bayesian networks of each relation into a single tree. We can also see that correlated sampling is relatively slow method, although its accuracy is competitive as shown in the previous subsection. MSCN is the slowest method overall in our benchmark. This may be attributed to the fact that we implemented it from scratch because no implementation was provided by its authors, and therefore do not have the insights that they might have. We argue that even though our method is not as accurate as the method proposed by [45], it is much faster and is thus more likely to be used in practice. Naturally, we also have to take into account the amount of time it requires to build our method, as well as how much storage space it requires.

4.4   Construction Time and Space

The cost model uses metadata that is typically obtained when the database isn't being used. This is done in order not to compute it in real time during the query optimisation phase. This metadata has to be refreshed every so often in order for the cost model to use relevant figures. Typically, the metadata has to be refreshed when the underlying data distributions change significantly. For instance, if attributes become correlated when new data is inserted, then the metadata has to be refreshed to take this into account. Therefore, the amount of time it takes to collect the necessary information is rather important, as ideally we would like to refresh the metadata as often as possible. Additionally, any viable selectivity estimation method crucially has to make do with a little amount of storage space. Indeed, spatial complexity is a major reason why most methods proposed in the literature are not being used in practice. These two computational requirements highlight the dilemma that cost models have to face: they have to be accurate whilst running with a very low footprint. Most multidimensional methods that have been proposed are utterly useless when it comes to their performance in this regard.

Table 8. Computational requirements of the construction phase per method on the JOB workload

	Construction time	Storage size
PostgreSQL	5 s	12 KB
Sampling	7 s	276 MB
Correlated sampling	32 s	293 MB
MSCN	15 min 8 s	37 MB
Global BN	24 min 45 s	429 KB
Independent BN	55 s	217 KB
Linked BN $k = 1$	2 min 3 s	322 KB
Linked BN $k = 2$	2 min 8 s	464 KB

Table 8 summarises the computational requirements of the methods we compared. The results explain why PostgreSQL – and most database engines for that matter – stick to using simplistic methods. Indeed, in our measurements PostgreSQL is both the fastest method as well the lightest one. PostgreSQL's cost model makes many simplifying assumptions and thus only has to build and store one-dimensional histograms, which can be done extremely rapidly. The sampling methods are quite fast in comparison with the methods based on

Bayesian networks. This isn't surprising, as they only require sampling the relations and then storing the samples. Indeed, most of the building time involves persisting the samples on the disk. On the other hand, sampling methods require a relatively large amount of space because they do not apply any summarising whatsoever (note that their storage size are given in terms of megabytes, not kilobytes). Correlated sampling takes more time than basic sampling because it has to scan primary and foreign keys in order to avoid the empty join problem. MSCN construction time is moderate, and naturally depends on the amount of data it is trained on. In this case we trained it for 20 epochs of stochastic gradient descent.

All of the methods based on Bayesian networks take more time to build than the two sampling methods, which is as expected. They make up in storage requirements and in inference time. The global Bayesian network takes a very large amount of time to build, which is in accordance with the results from [45]. In comparison, our method is much faster. This is a logical consequence of the fact that we only build one Bayesian network per relation. Additionally, each Bayesian network has a tree topology, which means that each conditional probability distribution we need to store is a two-way table. The sudden jump in building time between $k = 0$ and $k = 1$ is due to the need to compute joins when $k > 0$. However, note that the jump is much smaller between $k = 1$ and $k = 2$. The reason is that the joins don't have to be repeated for each additional attribute included in every parent Bayesian network.

5   Conclusion

During the query optimisation phase, a cost model is invoked by the query optimiser to estimate the cost of query execution plans. In this context, the selectivity of operators is a crucial input to the cost model [30]. Inaccurate selectivity estimates lead to bad cost estimates which in turn have a negative impact on the overall running time of a query. Moreover, errors in selectivity estimation grow exponentially throughout a query execution plan [22]. Selectivity estimation is still an open research problem, even though many proposals have been made. This is down to the fact that the requirements in terms of computational resources are extremely tight, and one thus has to compromise between accuracy and efficiency.

Our method is based on Bayesian networks, which are a promising way to solve the aforementioned compromise. Although the use of Bayesian networks for selectivity estimation isn't new, previous propositions entail a prohibitive building cost and inference time. In order to address these issues, we extend the work of [17] to include the measurement of dependencies between attributes of different relations. We show how we can soften the relational independence assumption without requiring an inordinate amount of computational resources. We validate our method by comparing it with other methods on an extensive workload derived from the JOB [30] and the TPC-DS [40] benchmarks. Our results show that our method is only slightly less accurate than the global Bayesian network from [45], whilst being an order of magnitude less costly to build and execute. Additionally, our method is more accurate than join-aware sampling, whilst requiring significantly less storage and computational requirements. In comparison with other methods which make more simplifying assumptions, our method is notably more accurate, whilst offering very reasonable guarantees in terms of computational time and space. In future work, we wish to extend our method to accommodate for specific operators such as GROUP BYs, as well as verify the benefits of our method in terms of overall query response time as perceived by a query issuer.

6 Appendix

6.1 Preliminary works

The following is an unabbreviated version of the subsection on our our preliminary subsection. It contains additional examples that help to get a better understanding, but that were considered too lengthy to be part of the main article.

In [17], we developed a methodology for constructing Bayesian networks to model the distribution of attribute values inside each relation of a database. Once the Bayesian networks are constructed, we used to produce selectivity estimates by converting a logical operator tree at hand into a probabilistic formula of sums and products. In what follows, we will give an overview of Bayesian networks. We will also explain the compromises we made in order to produce a method that is both reasonably accurate as well as efficient.

A Bayesian network is a probabilistic model. As such, it is used for approximating the probability distribution of a dataset. The particularity of a Bayesian network is that it uses a directed acyclic graph (DAG) in order to do so. The graph contains one node per variable, whilst each directed edge represents a conditional dependency between two variables. For instance, if nodes A and B are connected with an edge that points from A to B, then this stands for the conditional distribution $P(B \mid A)$. A Bayesian network is a product of many such conditional dependencies, which formally is:

$$P(X_1, \ldots, X_n) \simeq \prod_{X_i \in \mathcal{X}} P(X_i \mid Parents(X_i)) \qquad (18)$$

The term, $P(X_1, \ldots, X_n)$ is the probability distribution over the entire set of attributes $\{X_1, \ldots, X_n\}$. Meanwhile, $Parents(X_i)$ stands for the attributes that condition the value of X_i. The distribution $P(X_i \mid Parents(X_i))$ is thus the conditional distribution of attribute X_i's value. In practice, the full distribution is inordinately large, and is unknown to us. However, the total of the sizes of the conditional distributions $P(X_i \mid Parents(X_i))$ is much smaller. Indeed, for discrete attributes, each conditional distribution is a $(p + 1)$-way table, where p is the number of parents $|Parent(X_i)|$. If an attribute *hair* is conditioned by a single other attribute *nationality*, then that conditional relationship can be stored in a two-way table, as shown in Table 9.

Table 9. Conditional distribution $P(hair \mid nationality)$

	Blond	Brown	Dark
American	0.3	0.4	0.3
Japanese	0.05	0.1	0.85
Swedish	0.7	0.2	0.1

Meanwhile, if the *nationality* attribute is not conditioned by any other attribute, then we can represent with a one-dimensional distribution, which is represented in Table 10.

Table 10. Distribution $P(nationality)$

American	Japanese	Swedish
0.2	0.5	0.3

Assume the cost model is asked by the query optimiser to determine the fraction of tuples where *hair* equals *"Blond"* and *nationality* equals *"Swedish"*. This can be obtained by applying Bayes' rule:

$$
\begin{aligned}
P(hair = Blond, nationality = Swedish) &= P(Blond \mid Swedish) \times P(Swedish) \\
&= 0.7 \times 0.3 \\
&= 0.21
\end{aligned}
$$

(19)

Note that we do not have directly access to the distribution of the *hair* attribute. Indeed we only have the conditional distribution $P(hair \mid nationality)$. However, we can obtain $P(hair)$ by marginalising over the *nationality* attribute. For instance, if we want to obtain the fraction of tuples where *hair* equals *"Blond"*:

$$
\begin{aligned}
P(hair = Blond) &= \sum_{nationality} P(Blond, nationality) \\
&= \sum_{nationality} P(Blond \mid nationality) \times P(nationality) \\
&= \underbrace{0.3 \times 0.2}_{P(Blond, American)} + \underbrace{0.05 \times 0.5}_{P(Blond, Japanese)} + \underbrace{0.7 \times 0.3}_{P(Blond, Swedish)} \\
&= 0.295
\end{aligned}
$$

(20)

With a Bayesian network, we are thus able to answer any selectivity estimation problem by converting a logical query into a mathematical formula following standard rules of probability [24]. Note, however, that a Bayesian network is necessarily an approximation of the full probability distribution because it makes assumptions about the generating process of the data. Finding the right graph structure of a Bayesian network is called *structure learning* [24]. This is usually done using a scoring function, which estimates the amount of information memorised by a network with a given structure. The time required to run an exhaustive search which maximises the scoring function is super-exponential with the number of variables [12]. Approximate search methods as well as integer programming solutions have been proposed [3], but they still require a large amount of time to run and have brittle performance guarantees. In our work in [17], we proposed to use the *Chow-Liu* algorithm [11]. This algorithm has the property of finding the best tree structure where nodes are restricted to have at most one parent. The obtained tree is the best in the sense of maximum likelihood estimation. In other words, it is the tree that memorises the most the given data. This is an important property, because for the purpose of selectivity estimation we are not interested in having a model that generalises well, but rather one that is good at memorising the data that it is shown. This is explained in further details in Sect. 4.1 of [16]. In addition to this property, the Chow-Liu algorithm only runs in $\mathcal{O}(p^2)$ time, where p is the number of variables, and is simple to implement. It works by first computing the *mutual information* between each pair of variables, which is defined as so:

$$MI(X_i, X_j) = \sum_{x_i \in X_i} \sum_{x_j \in X_j} P(x_i, x_j) \times \log(\frac{P(x_i, x_j)}{P(x_i)P(x_j)}) \qquad (21)$$

The mutual information can be seen as the strength of the relation between two variables, whether it be linear or not. The distribution $P(X_i, X_j)$ contains the occurrence counts of each pair (x_i, x_j) in a relation R. It can be obtained with a `SELECT COUNT(*) FROM R GROUP BY` X_i, X_j statement in `SQL`. The distributions $P(X_i)$ and $P(X_j)$ can be obtained by marginalising over $P(X_i, X_j)$ with respect to the other attribute. Once the mutual information for each pair of attributes is computed, they are organised into a fully connected weighted graph, as shown in Fig. 7:

The next step is to find the *maximum spanning tree* (MST) of the graph, which is the spanning tree whose sum of edge weights is maximal. A spanning tree is a subset of $p-1$ edges that forms a tree. Finding the maximum spanning tree can be done in $\mathcal{O}(p \log(p))$ time, for example, by using Kruskal's algorithm [28]. The maximum spanning tree is then turned into a directed graph by choosing a root attribute. The choice of the root attribute does not matter because mutual information is symmetric. The result of applying this procedure to the graph from Fig. 7 is shown in Fig. 8.

Once the structure of a Bayesian network has been decided upon, it can be used to answer probabilistic queries. This is usually referred to as *inference*. Inference for Bayesian networks is known to be NP-hard [12]. Exact inference as

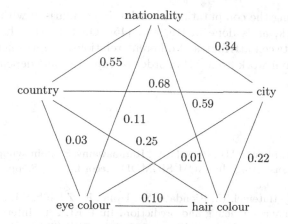

Fig. 7. Mutual information amounts for five attributes

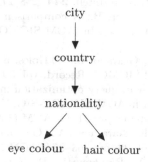

Fig. 8. Maximum spanning tree of Fig. 7

well as approximate methods have been proposed. The most basic inference algorithm is called the *variable elimination* algorithm [13] and is an exact inference algorithm. It works by explicitly writing the inference equation defined by the Bayesian network's structures, and then moving the sum and product operators around in order to "eliminate" repetitive calculations. In the case of trees, this can be done in linear time, which is the main reason why we constrained our Bayesian networks to have tree topologies in [17]. Many other inference algorithms exist; this includes belief propagation (used in [45]), linear programming, sampling methods, and variational inference. However, our experiments indicated that all of these were much slower than the variable elimination algorithm in the case of trees. Our inference process can further be accelerated by identifying branches of the Bayesian network that do not pertain to a particular query. This identification process is called the *Steiner tree problem* [19].

In [17], we proposed a simple method which consists in building one Bayesian network per relation. We used the Bayesian networks to estimate the selectivity of queries inside their respectful relations. On the one hand, this has the benefit of greatly reducing the computational burden in comparison with a single large Bayesian network, as is done in [16] and [45]. On the other hand, it ignores dependencies between attributes of different relations. We will now discuss how we can improve our work from [17] in order to capture some dependencies across relations.

References

1. Acharya, S., Gibbons, P.B., Poosala, V., Ramaswamy, S.: Join synopses for approximate query answering. In: ACM SIGMOD Record, vol. 28, pp. 275–286. ACM (1999)
2. Akdere, M., Çetintemel, U., Riondato, M., Upfal, E., Zdonik, S.B.: Learning-based query performance modeling and prediction. In: IEEE 28th International Conference on Data Engineering (ICDE), pp. 390–401. IEEE (2012)
3. Bartlett, M., Cussens, J.: Integer linear programming for the Bayesian network structure learning problem. Artif. Intell. **244**, 258–271 (2017)
4. Blohsfeld, B., Korus, D., Seeger, B.: A comparison of selectivity estimators for range queries on metric attributes. In: ACM SIGMOD Record, vol. 28, pp. 239–250. ACM (1999)
5. Bruno, N., Chaudhuri, S., Gravano, L.: STHoles: a multidimensional workload-aware histogram. In: ACM SIGMOD Record, vol. 30, pp. 211–222. ACM (2001)
6. Chaudhuri, S.: An overview of query optimization in relational systems. In: Proceedings of the Seventeenth ACM SIGACT-SIGMOD-SIGART Symposium on Principles of Database Systems, pp. 34–43. ACM (1998)
7. Chaudhuri, S., Motwani, R., Narasayya, V.: On random sampling over joins. In: ACM SIGMOD Record, vol. 28, pp. 263–274. ACM (1999)
8. Chaudhuri, S., Narasayya, V., Ramamurthy, R.: Exact cardinality query optimization for optimizer testing. Proc. VLDB Endowment **2**(1), 994–1005 (2009)
9. Chen, C.M., Roussopoulos, N.: Adaptive selectivity estimation using query feedback, vol. 23. ACM (1994)
10. Chen, Y., Yi, K.: Two-level sampling for join size estimation. In: Proceedings of the 2017 ACM International Conference on Management of Data, pp. 759–774. ACM (2017)
11. Chow, C., Liu, C.: Approximating discrete probability distributions with dependence trees. IEEE Trans. Inf. Theory **14**(3), 462–467 (1968)
12. Cooper, G.F.: The computational complexity of probabilistic inference using Bayesian belief networks. Artif. Intell. **42**(2–3), 393–405 (1990)
13. Cowell, R.G., Dawid, P., Lauritzen, S.L., Spiegelhalter, D.J.: Probabilistic Networks and Expert Systems: Exact Computational Methods for Bayesian Networks. Springer, New York (2006). https://doi.org/10.1007/b97670
14. Deshpande, A., Garofalakis, M., Rastogi, R.: Independence is good: dependency-based histogram synopses for high-dimensional data. ACM SIGMOD Record **30**(2), 199–210 (2001)
15. Dutt, A., Wang, C., Nazi, A., Kandula, S., Narasayya, V., Chaudhuri, S.: Selectivity estimation for range predicates using lightweight models. Proc. VLDB Endowment **12**(9), 1044–1057 (2019)
16. Getoor, L., Taskar, B., Koller, D.: Selectivity estimation using probabilistic models. In: ACM SIGMOD Record, vol. 30, pp. 461–472. ACM (2001)

17. Halford, M., Saint-Pierre, P., Morvan, F.: An approach based on Bayesian networks for query selectivity estimation. In: Li, G., Yang, J., Gama, J., Natwichai, J., Tong, Y. (eds.) DASFAA 2019. LNCS, vol. 11447, pp. 3–19. Springer, Cham (2019). https://doi.org/10.1007/978-3-030-18579-4_1

18. Heimel, M., Kiefer, M., Markl, V.: Self-tuning, GPU-accelerated kernel density models for multidimensional selectivity estimation. In: Proceedings of the 2015 ACM SIGMOD International Conference on Management of Data, pp. 1477–1492. ACM (2015)

19. Hwang, F.K., Richards, D.S., Winter, P.: The Steiner Tree Problem, vol. 53. Elsevier, North-Holland (1992)

20. Ioannidis, Y.: The history of histograms (abridged). In: Proceedings 2003 VLDB Conference, pp. 19–30. Elsevier (2003)

21. Ioannidis, Y.E.: Query optimization. ACM Comput. Surv. (CSUR) **28**(1), 121–123 (1996)

22. Ioannidis, Y.E., Christodoulakis, S.: On the propagation of errors in the size of join results, vol. 20. ACM (1991)

23. Ivanov, O., Bartunov, S.: Adaptive cardinality estimation. arXiv preprint arXiv:1711.08330 (2017)

24. Jensen, F.V., et al.: An Introduction to Bayesian Networks, vol. 210. UCL press, London (1996)

25. Kipf, A., Kipf, T., Radke, B., Leis, V., Boncz, P., Kemper, A.: Learned cardinalities: estimating correlated joins with deep learning. arXiv preprint arXiv:1809.00677 (2018)

26. Kipf, A., et al.: Estimating cardinalities with deep sketches. In: Proceedings of the 2019 International Conference on Management of Data, pp. 1937–1940 (2019)

27. Kooi, R.P.: The optimization of queries in relational databases (1981)

28. Kruskal, J.B.: On the shortest spanning subtree of a graph and the traveling salesman problem. Proc. Am. Math. Soc. **7**(1), 48–50 (1956)

29. Kschischang, F.R., Frey, B.J., Loeliger, H.A., et al.: Factor graphs and the sum-product algorithm. IEEE Trans. Inf. Theory **47**(2), 498–519 (2001)

30. Leis, V., Gubichev, A., Mirchev, A., Boncz, P., Kemper, A., Neumann, T.: How good are query optimizers, really? Proc. VLDB Endowment **9**(3), 204–215 (2015)

31. Leis, V., et al.: Query optimization through the looking glass, and what we found running the join order benchmark. VLDB J. **27**, 643–668 (2018)

32. Leis, V., Radke, B., Gubichev, A., Kemper, A., Neumann, T.: Cardinality estimation done right: index-based join sampling. In: CIDR (2017)

33. Liu, H., Xu, M., Yu, Z., Corvinelli, V., Zuzarte, C.: Cardinality estimation using neural networks. In: Proceedings of the 25th Annual International Conference on Computer Science and Software Engineering, pp. 53–59. IBM Corp. (2015)

34. Markl, V., Haas, P.J., Kutsch, M., Megiddo, N., Srivastava, U., Tran, T.M.: Consistent selectivity estimation via maximum entropy. VLDB J. **16**(1), 55–76 (2007)

35. Matias, Y., Vitter, J.S., Wang, M.: Wavelet-based histograms for selectivity estimation. In: ACM SIGMOD Record, vol. 27, pp. 448–459. ACM (1998)

36. Moerkotte, G., Neumann, T., Steidl, G.: Preventing bad plans by bounding the impact of cardinality estimation errors. Proc. VLDB Endowment **2**(1), 982–993 (2009)

37. Müller, M., Moerkotte, G., Kolb, O.: Improved selectivity estimation by combining knowledge from sampling and synopses. Proc. VLDB Endowment **11**(9), 1016–1028 (2018)

38. Muralikrishna, M., DeWitt, D.J.: Equi-depth multidimensional histograms. In: ACM SIGMOD Record, vol. 17, pp. 28–36. ACM (1988)

39. Olken, F., Rotem, D.: Simple random sampling from relational databases (1986)
40. Poess, M., Smith, B., Kollar, L., Larson, P.: TPC-DS, taking decision support benchmarking to the next level. In: Proceedings of the 2002 ACM SIGMOD International Conference on Management of Data, pp. 582–587 (2002)
41. Poosala, V., Haas, P.J., Ioannidis, Y.E., Shekita, E.J.: Improved histograms for selectivity estimation of range predicates. In: ACM SIGMOD Record, vol. 25, pp. 294–305. ACM (1996)
42. Poosala, V., Ioannidis, Y.E.: Selectivity estimation without the attribute value independence assumption. VLDB **97**, 486–495 (1997)
43. Selinger, P.G., Astrahan, M.M., Chamberlin, D.D., Lorie, R.A., Price, T.G.: Access path selection in a relational database management system. In: Proceedings of the 1979 ACM SIGMOD International Conference on Management of Data, pp. 23–34. ACM (1979)
44. Stillger, M., Lohman, G.M., Markl, V., Kandil, M.: Leo-DB2's learning optimizer. VLDB **1**, 19–28 (2001)
45. Tzoumas, K., Deshpande, A., Jensen, C.S.: Lightweight graphical models for selectivity estimation without independence assumptions. Proc. VLDB Endowment **4**(11), 852–863 (2011)
46. Van Aken, D., Pavlo, A., Gordon, G.J., Zhang, B.: Automatic database management system tuning through large-scale machine learning. In: Proceedings of the 2017 ACM International Conference on Management of Data, pp. 1009–1024. ACM (2017)
47. Vengerov, D., Menck, A.C., Zait, M., Chakkappen, S.P.: Join size estimation subject to filter conditions. Proc. VLDB Endowment **8**(12), 1530–1541 (2015)
48. Wu, W., Chi, Y., Zhu, S., Tatemura, J., Hacigümüs, H., Naughton, J.F.: Predicting query execution time: are optimizer cost models really unusable? In: IEEE 29th International Conference on Data Engineering (ICDE), pp. 1081–1092. IEEE (2013)
49. Yin, S., Hameurlain, A., Morvan, F.: SLA definition for multi-tenant DBMS and its impact on query optimization. IEEE Trans. Knowl. Data Eng. **30**, 2213–2226 (2018)

Author Index

Printed in the United States
By Bookmasters